Plastic Circularity

Data Development, Flow Analysis, and Value Chain Coordination

Plastic Circularity

Data Development, Flow Analysis, and Value Chain Coordination

Paul T. Mativenga
Department of Mechanical and Aerospace Engineering, School of Engineering, The University of Manchester, Manchester, United Kingdom; Postgraduate School of Engineering Management, University of Johannesburg, Johannesburg, South Africa

Kunle Ibukun Olatayo
Postgraduate School of Engineering Management, University of Johannesburg, Johannesburg, South Africa

Annlizé L. Marnewick
Postgraduate School of Engineering Management, University of Johannesburg, Johannesburg, South Africa

ELSEVIER

Elsevier
Radarweg 29, PO Box 211, 1000 AE Amsterdam, Netherlands
125 London Wall, London EC2Y 5AS, United Kingdom
50 Hampshire Street, 5th Floor, Cambridge, MA 02139, United States

ISBN: 978-0-443-40667-6

For Information on all Elsevier publications
visit our website at https://www.elsevier.com/books-and-journals

Publisher: Peter Llewellyn
Acquisitions Editor: Pamela Sadhukhan
Editorial Project Manager: Juwairiya Ali
Production Project Manager: Gomathi Sugumar
Cover Designer: Mark Rogers

Typeset by MPS Limited, Chennai, India

Contents

Preface

Plastic pollution in our waters, on land and even our food systems is a major and growing threat to biodiversity and human health. At the same time, plastic materials and products remain indispensable, delivering immense value across sectors such as healthcare, in our hospitals, food preservation and beyond. This paradox highlights a critical tension between dependence and damage.

To crack this challenge, transitioning to a circular economy for plastics has become an urgent and global priority. The circular economy for plastics is achieved through three core principles: eliminating waste and pollution, circulating products and materials at their highest value and regenerating nature.

This book offers fundamental knowledge on the flow of plastic materials, considering nations as system boundaries, and explores innovative, evidence-based pathways for building circular systems.

Drawing on peer-reviewed science and insights from international collaboration, this timely work invites citizens, governments and institutions alike to engage with one of the defining environmental challenges and opportunities of our time.

Acknowledgements

The authors are grateful to the earlier funding by the Engineering and Physical Science Research Council (EPSRC), United Kingdom, on a project in "RE3 – Rethinking Resources and Recycling" (Ref: EP/S025200/1).

Chapter 1

Introduction

1.1 Book introduction

1.1.1 Circular economy: nature is circular

Days, weeks, months and years move in a steady, repeating rhythm and the water cycle continually renews the planet's supply, sustaining life through constant regeneration. Yet our global economies have long operated in stark contrast – extracting resources, using them and discarding the waste in a linear 'take-make-dispose' pattern. Today, this linear model is under intense pressure as resource depletion, scarcity, pollution and climate change expose its limits and demand a new way forward.

The circular economy encourages a shift from the traditional linear model of resource consumption based on take-make-use-dispose to one based on eliminating waste and enabling resource restoration and regeneration. The circular economy is delivered by realising options higher up the waste hierarchy, such as reduce, reuse, recycle and recover, within the stages of production, distribution, use, consumption and end-of-life. Success enables the reduction of raw material extraction, reduced resource use, extension of product lifetime, reprocessing of the product at the end-of-life, minimisation of waste and regeneration of the natural system. The circular economy enables value retention in materials and products, supports material security, enables efficient use of resources, rethinks waste as a resource and can be optimised to reduce environmental impact. All these deliverables provide a strong incentive to adopt this sustainable model.

A major misconception about the circular economy is that it is merely waste management. The model is much more than that, as circular economy strategies are applicable or can be embedded at every stage of the plastics value chain system. These stages include design, production, distribution, consumption and end-of-life.

While the case for a circular economy is very strong and compelling and the transition to a circular economy is urgent, progress has been slow globally. Transitioning to circularity demands systemic change. Transforming the whole value chain system requires effort from stakeholders, including individuals, businesses, universities, innovation centres and governments.

The transition towards a circular economy is critical for several sustainability benefits, economically, socially and environmentally, as presented in Table 1.1.

Plastic Circularity. DOI: https://doi.org/10.1016/B978-0-443-40667-6.00013-5

TABLE 1.1 Economic, social and environmental sustainability benefits of the circular economy.

	Benefits	Description
Economic	Cost savings	• Less frequency and replacement cost when products are durable, reusable or repairable • Reduced cost of waste disposal • Material reuse and re-manufacturing reduce operational costs, saving on new materials and waste management • Circular practices reduce water and energy consumption, saving cost
	Resource supply security	• Reduced dependence on virgin material inputs reduces vulnerability to resource scarcity • Better supply chain stability and resilience • Lowers risks associated with supply price volatility
	New circular business opportunities	• New businesses and markets are created across the value chain, from the product design stage
	Job creation	• More employment opportunities from new businesses across the circular economy value chain
	Economic growth	• Circularity solutions are stimulus to innovation across different sectors of the economy, creating new economic outputs • Innovative circular business models result in sustainable economic growth
Social	Improved livelihood and public health	• Economic growth promotes a higher standard of living and increased public well-being • Sustainable cleaner environment results in improved public health and living conditions
	Local empowerment	• Circular initiatives promote localisation of production • More community control over their resources and environment • The economic opportunities created reduce inequality and marginalisation • Promotion of social cohesion

(*Continued*)

TABLE 1.1 (Continued)

	Benefits	Description
	Compliance and improved corporate image	• Corporate image and reputation are improved and strengthened • More customers and partners with shared values are attracted and connected
Environmental	Resource conservation and efficiency	• Reduced demand for virgin materials helps conserve natural ecosystems and biodiversity • Sustainable product design increases material efficiency, which reduces excess consumption
	Ecosystem regeneration	• Resource efficiency and conservation limit landscape, habitat and biodiversity loss or disruption • Regenerative farming and water recycling restore environmental health (soil fertility, biodiversity and ecosystem resilience)
	Limited waste and pollution	• By extending the life cycle of products and eliminating unnecessary materials, waste generation decreases significantly
	Limited emissions	• Less harmful carbon and greenhouse gas emissions are released into the air, water and soil

1.1.2 Principles of a circular economy

The Ellen Macarthur Foundation identified three fundamental principles that drive the circular economy, by design:

- Elimination of waste and pollution
- Circulation of products and materials in use (at their highest value)
- Regeneration of nature

The first principle of elimination of waste forces stakeholders to identify and eliminate waste at every stage of the value chain. The relevant stages can start from the processing of plastic materials and move through the production of plastic products, use of plastic materials and products and end-of-life of plastic materials and products. Identified waste is regarded as a potential resource and not something to be disposed of. In addition, extending the lifespan of products helps eliminate the rate of waste generation.

The principle of circulating products and materials (at their highest value) promotes keeping materials in use either as a product or alternatively as components or raw materials. This is supported by reuse, re-manufacturing, re-purposing and recycling.

The principle of regeneration of nature promotes natural processes and enables nature to flourish. It supports environmental impact minimisation, biodiversity conservation and ecological restoration. Designing sustainability into products and processes from the design phase helps to limit waste generation and conserve resources.

The circular economy can follow the biological and technical cycles. The features of the biological cycle include farming/collection, biochemical feedstock, regeneration, anaerobic digestion and extraction of biochemical feedstock. Nutrients from biodegradable materials are returned to the Earth for the regeneration of nature. The technical cycle keeps materials and products in circulation by applying processes such as sharing, maintaining/prolonging, reusing/redistributing, refurbishing/re-manufacturing and recycling. The processes of sharing, prolonging and reusing will extend a product's lifespan and make use of resources more efficiently than re-manufacturing and recycling. Re-manufacturing/refurbishment will demand replacing some components of a product, and recycling requires the return of the original product to raw materials.

1.1.3 How is success measured? Metrics of a circular economy

Making progress regarding the circular economy necessitates measuring performance, identifying priority areas or hot-spots for action and setting new targets.

Metrics are indicators or functions of measurement to calculate performance or progress made. They are basically data for insight. The need for data-driven informed decisions in implementing or adopting circularity initiatives across a product value chain is driving the development of circularity metrics. Circularity metrics act as the operational backbone for evaluating how effective, scalable and impactful circular initiatives are across businesses, sectors, industries and government organisations, as well as locally, regionally and nationally.

They are imperative in guiding the transition to the circular economy and assessing impact. The variation of metrics to calculate the circularity of a system has increased over the years. Several circularity tools or models have been developed to guide the measurement of performance and progress on circularity.

More remarkably, the introduction of ISO 59020 signifies a major landmark in the evolution of circularity assessment standardisation. It is an international standard that provides a structured and compliant framework for quantifying the circular performance of materials, products, businesses, organisations, sectors and nations. The standard defines and differentiates principal and secondary indicators and describes the systematic methods and procedures for collecting, analysing and interpreting data. However, there are limitations to circular metrics, particularly data availability and quality. Also, metrics must be characterised by consistency, wide acceptability, clarity and comprehensibility for it to support the shift to circularity successfully.

The material circularity index (MCI) has performance values that range from 0 to 1. The value 0 for MCI indicates that the product was fully manufactured from virgin material and is totally disposed of at the product's end-of-life. This also highlights that the product's utility is at a low level. A value of 1 signifies that the product is totally circular, totally making use of only reused and recycled resources and that there was no disposal of material at the product's end-of-life.

Metrics such as carbon footprint, water footprint and material scarcity are supporting indicators of the primary metrics. The combination helps make all-inclusive decisions regarding the circularity of a product or system. Carbon footprint relates to activities such as the energy consumed and transportation associated with material and product production, quantifying the carbon and other greenhouse gas emissions from these activities. Water footprints represent the total litres used during production. Material scarcity describes the relative availability or abundance of a material or the amount of material used for a product. It encourages the efficient use of material for products.

1.1.4 Vision for a circular plastic economy

Plastics are ubiquitous and everywhere. Due to their unique and outstanding properties, plastics have been put to extensive use and numerous functions in a range of products and applications. The advancement of the properties of plastic materials (strength, mouldability, durability, light weight, liquid impermeability, corrosion resistance and thermal insulation) has made them more applicable in diverse sectors of the economy, such as health, safety, food, clothing, vehicles and furniture, thereby transforming the modern world.

Regrettably, despite the successful application of plastics and the numerous benefits they have brought to the economy and society, they can be problematic due to their non-decomposing nature. Several of the desirable qualities of plastic equally make the material highly resistant to environmental bio-degradation. This non-decomposing nature of plastics results in the accumulation of the material in the ecosystem.

The sustained accumulation of plastics leads to waste challenges for the environment and biodiversity. Plastic pollution is among the biggest recent environmental challenges and is now a common occurrence in the ecosystem and in every aspect of human life: the soil, the ocean, the atmosphere, our drinking water, food and in our bodies. This is because most plastics produced are being mismanaged after their end-of-life, ending up in landfills or in the natural environment as plastic pollution.

At this period in history, the relationship society has with plastics is evolving. All stakeholders across the plastics value chain, from the government to the plastics industry and general society, largely want to achieve a singular objective: to limit plastic waste generation and make the environment more sustainable. This objective presents an opportunity for countries to promote more circular business models, develop sustainability infrastructure and embed a culture that prioritises the reduction, reuse and recycling of plastics.

The vision of achieving the transition to a circular plastic economy requires rethinking and redesigning entire plastics value chains, rethinking plastic use and ensuring that where plastic is used, the value of plastic materials and products is retained within the economy, and plastic waste and associated environmental damage are minimised. This requires implementing transformative changes in policy, industry practices and processes, as well as consumer behaviour.

The vision for plastic circularity requires circularity to be adopted and integrated at every stage of the plastics value chain. This necessitates a value chain approach that considers the upstream, midstream and downstream stages of plastic materials and products.

Circular intervention at the upstream stage of the plastics value chains begins with the reduced or efficient use of non-toxic virgin plastics and other sustainable additive inputs, as well as the adoption of sustainable product design and sustainable production processes for circularity. Designing plastic products based on circular and sustainable design fosters the reduction or elimination of waste and retention of plastics within the economy, and plastic value is preserved.

At the midstream stage of the plastics value chain, implementing circularity involves distributing plastic products through the least polluting means, and reusing, re-purposing, repairing and re-manufacturing the product for an extended lifespan.

The downstream circularity intervention in the plastics value chain targets the end-of-life of the plastic product after consumption. The emphasis is on the return of the product and its parts back to the economy through separate collection, recovery and recycling of the corresponding plastic waste to close the loop.

1.1.5 Structure of the book

This book presents a framework and insights for transforming plastic material and product value chains into an integrated circular economy. It is informed by cutting-edge research and practical tools to help overcome the major challenge of plastic waste and realise the opportunities for circular models. It focuses on accelerating the transition to a circular economy for plastics, while maximising value retention by promoting options higher up the waste management hierarchy. The scope of the book covers tools for modelling plastic material flows within a defined system, envisioning circularity options that retain maximum material value, developing metrics to evaluate such systems and methodologies for transitioning to higher and desired levels of circularity. The vision is to enable a rapid transition to tighter circularity loops.

The concept of the book involves two major parts regarding plastic circularity – the background and the roadmap – as illustrated by Fig. 1.1.

The background contents flow from discussing the basics of plastics in the economy to plastics in the circular economy, plastic waste as an integral aspect of the circular economy, the consequences of plastic waste and sustainability, and the management of plastic waste regarding the waste hierarchy systems. The roadmap contents explore critical factors to achieving plastic circularity, including data and plastic flow models for improved planning and strategy, as well as the coordination of the plastics value chain by stakeholders for improved and increased implementation.

The book contains nine chapters: Chapter 2. Plastics in the economy and the value chain. Chapter 3. Plastic waste and the waste management hierarchy. Chapter 4. Impact and sustainability of plastic. Chapter 5. Data building in the plastics value chain. Chapter 6. Modelling of plastic material flows for sustainable decision-making. Chapter 7. Participation and dynamics of stakeholders in the value chain. Chapter 8. Progress towards plastic circularity. Chapter 9. Transition towards a world-class system for plastic circularity: Performance and maturity assessment. Chapter 10. Case studies on the country transition to a circular plastic economy. Chapter 11. Conclusion.

1.1.5.1 Chapter 2: plastics economy and the value chain

The discussions in this chapter cover the contributions of plastics to the economy, the value chain and the different stages, the vision of a circular economy, as well as the various opportunities and challenges of the plastic system with respect to the integration of circularity across the value chain. It is shown that while plastic has brought so much value, its life cycle, which covers production, use and end-of-life, comes with several problems and associated negative impacts where it is unsustainably consumed and the waste is poorly managed. The vision of a circular economy and its application to the plastic system creates an economy where plastic waste is controlled or eliminated by promoting sustainable plastic product design and manufacture, intensive use, reuse, repair, refurbishment and recycling.

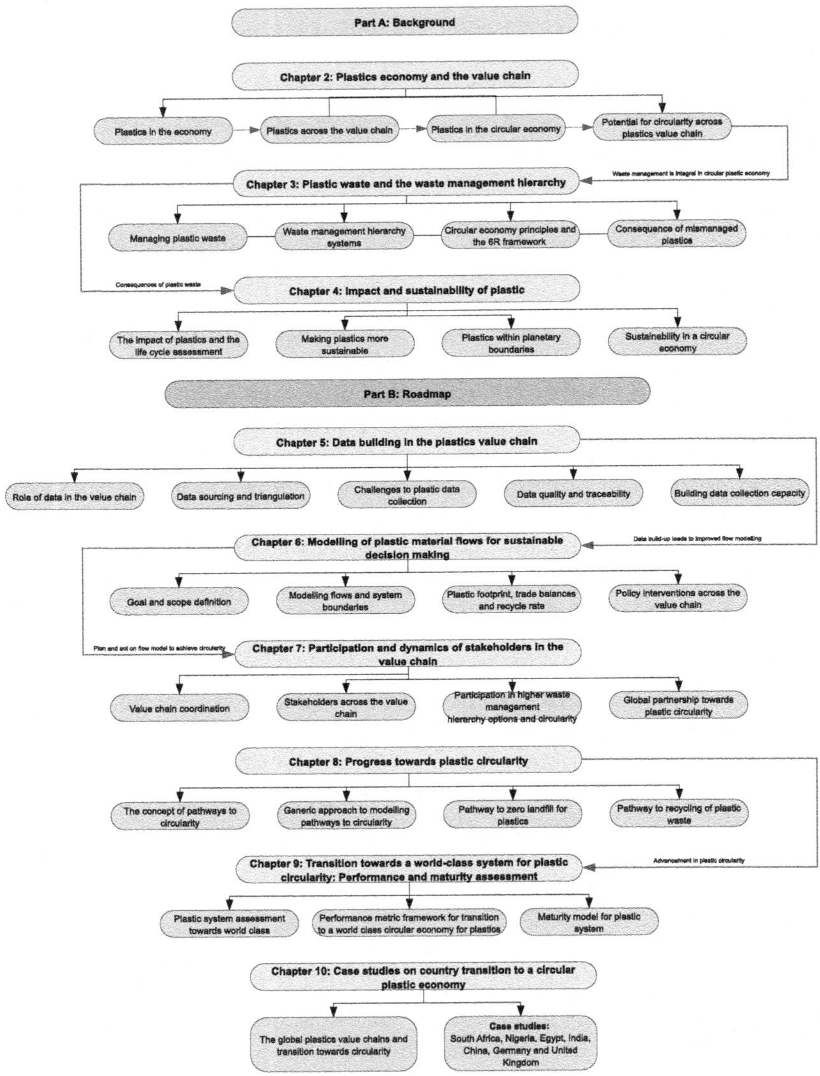

FIGURE 1.1 **Visual of the book structure.**

1.1.5.2 Chapter 3: plastic waste and the waste management hierarchy

In this chapter, the management of plastics and plastic waste is examined by exploring the waste management hierarchy systems, the link between the waste hierarchy options and the circular economy principles, as well as the consequences of mismanaged plastics. Advancing the principles of the circular economy for plastics requires a focus on the *redesign, reduce, re-manufacture, reuse, recycle* and *recovery options*, in that *order and priority*. Broader system transformation frameworks are required to promote and enable a shift towards the options higher up the waste management hierarchy.

1.1.5.3 Chapter 4: impact and sustainability of plastic

This chapter deals with the impact and sustainability of plastics in society. The impacts of the different phases of the life cycle and how to assess them using life cycle assessment are discussed. Sustainability is introduced within planetary boundaries and a circular economy. Plastics, from production to use and end-of-life, have adverse impacts on the environment. Realising plastic sustainability and operating within planetary boundaries requires integrating the circular economy principles.

1.1.5.4 Chapter 5: data building in the plastics value chain

The significance of data in the plastics value chain is examined in this chapter, covering the diverse sources of plastic data collection, the challenges in collecting data and measures for building data collection capacity across the plastics value chain. Data in the plastics value chain is crucial in assessing, planning, strategising, modelling and implementing solutions to reduce plastic waste and increase plastic circularity. A framework for building plastic data capacity within the value chain is proposed to address data gaps in the system, as well as the many challenges associated with data collection in the different value chain stages. A unified national registry for plastics would enable better coordination for efficient plastic data collection, reporting and tracking system performance.

1.1.5.5 Chapter 6: modelling of plastic material flows for sustainable decision-making

The modelling of plastic material flows is examined in this chapter, particularly the application of the material flow analysis (MFA) technique for modelling the mass flow and flow processes across the plastics value chain. The chapter covers system boundaries, the national plastic footprint, trade balance and utility in calculating the national plastic recycling rate. From production to consumption, trade, distribution and end-of-life management, plastics interrelate with the environment and the economy in multi-layered ways. Therefore the plastic material flow model provides valuable data and information for making informed decisions and policies for the advancement of circularity and sustainability of the plastic system and the environment.

1.1.5.6 Chapter 7: participation and dynamics of stakeholders in the value chain

This chapter covers the role of stakeholders in advancing a circular economy for plastics across the value chain. The plastics value chain is dynamic and multi-staged, with multiple stakeholders in every stage. These stakeholders have objectives which need to be coordinated. Well-coordinated plastics systems produce increased household recycling awareness and motivation, better waste handling and sorting, improved regulation enforcement and reduced expenses from cost sharing of facilities and infrastructure. For the plastics system to transition towards circularity, the effective participation of stakeholders in the value chain is crucial for success.

1.1.5.7 Chapter 8: progress towards plastic circularity

This chapter focuses on pathways to circularity of the plastic system. The transition to circularity is normally driven by a time-bound circularity target for a country or region. There are a few pathways that can be considered to achieve circularity targets. These targets may be related to reaching zero landfill, recycling, re-manufacturing and reuse rates by a defined period. To guide this transition, it is important to identify and assess the potential of different strategies to deliver to the transition pathway. This chapter presents an example of how to identify and evaluate strategies to deliver the circular economy transition.

1.1.5.8 Chapter 9: transition towards a world-class system for plastic circularity: performance and maturity assessment

This chapter covers how to improve a circular economy for plastics and transition to a world-class system. A maturity model is defined to assess the plastic material system's improvement and development, regardless of geographical location.

1.1.5.9 Chapter 10: case studies on country transition to a circular plastic economy

To gain insight into transitioning towards a circular economy, the progress of seven countries towards circularity in plastics is presented, considering local context. The plastic value chain systems of these countries are discussed, including efforts in advancing the circularity of plastics. The review of the systems provides an understanding of the plastics value chain for informed policy-making in improving circularity across the value chain stages and making the plastic industry more environmentally responsible.

Chapter 2

Plastics economy and the value chain

2.1 Plastics in the economy

Plastic is anywhere and everywhere. It is a significant material in modern societies, and practically every element of our daily lives involves the application of plastic in one way or another. It is the preferred material in a broad range of applications, for example, in health, safety, food, clothing, vehicles and furniture. Plastic material products are used in hospitals. Plastic enables the design and development of lighter vehicles. It is used to insulate electrical cables for maximum protection and efficiency, and in pipework and equipment for the supply of clean water for drinking. Plastic packaging prolongs the life of food, thus reducing wastage and food-related illness. Fig. 2.1 shows some examples of applications of plastic materials ranging from plastic bottles to containers, beads, toys, waste bins and furniture.

While plastics have been applied successfully, are prominent in the economy and have brought so much value to us, they also come with challenges and can be problematic. They typically end up in the environment and do not decompose. Some of the qualities that make plastics desirable for numerous applications are the very qualities that also make them highly resistant to environmental biodegradation. Regrettably, this is resulting in the accumulation of the material.

The continued accumulation of this non-decomposing material causes problems for the environment and biodiversity. Plastic pollution is one of the biggest environmental challenges in recent years and is now a common occurrence in the ecosystem and in every aspect of human life: the soil, the ocean, the atmosphere, our drinking water, food and in our bodies. Images and videos abound of beaches and oceans littered with plastic, as well as turtles, dolphins and seabirds ingesting or entangled in plastic. This scenario is expected to continue as plastic production, which has been projected to multiply in the future, continues to outstrip current waste management and recycling capacities.

Most plastics produced are being mismanaged and are not reused or recycled after their end-of-life, ending up in landfills or in the natural

Plastic Circularity. DOI: https://doi.org/10.1016/B978-0-443-40667-6.00003-2

FIGURE 2.1 Applications of plastic.
Photos showing the use of plastic. Photo credit: Kunle Olatayo.

environment as plastic pollution. Fig. 2.2 shows some typical environmental pollution challenges from plastic waste on land and in our waters.

Improving the circularity of plastic materials and products is an urgent priority and crucial in addressing the associated problems of plastic waste and its negative environmental impact. Tackling the scope and impacts of plastics on the environment requires action from stakeholders, both government and non-government entities.

2.1.1 What makes plastic unique and widely used?

Plastics have various characteristic properties. They are tough and strong, light- and chemical-resistant and can be biocompatible and easily worked. These and other properties make plastics unique and widely utilised compared to other materials. As a result, they have been put to extensive use and have numerous functions in a range of products and applications. Plastics are strong, mouldable, durable and lightweight. Other unique properties include liquid impermeability, corrosion resistance and thermal insulation (Bucknall, 2020). The advances in these desirable properties have increased the applicability of

FIGURE 2.2 Plastic pollution on land and water.
Photo showing plastic pollution on water and land. Photo credit: Kunle Olatayo.

plastic in different sectors, including transportation, the automotive industry, packaging, health, agriculture, textiles, furniture, electronics and construction. Depending on the type of plastic, applications can be single-use, or short- or long-term service. Single-use applications are common, such as packaging and shopping bags, whereas long-term use applications are common in construction and vehicle components.

2.1.2 Plastic is helping the food industry

The biggest application of plastic is packaging. The share of the global volume of packaging materials increased from 17% to 25% between 2000 and 2015, with a growth of 5% annually (World Economic Forum, 2016). Plastic is revolutionising the packaging industry and has immense benefits in the food industry. Fruit juices and drinks (carbonated and non-carbonated) are bottled in plastic packaging. The shelf life of food is extended, making vital health applications possible and enabling safer and lighter shipping, thus reducing transport costs and emissions. Plastics preserve food for an extended period by shielding it from physical damage, especially while being transported, as well

FIGURE 2.3 Plastic packaging for food and drinks.
Photos of plastic packing. Photo credit: Kunle Olatayo.

as biological and chemical damage. Food preserved in plastic packaging is protected against bacteria and less prone to causing infection (Bucknall, 2020). It stays fresher for longer, preserving the taste and nutritional value of food and leading to less waste. Considering that tonnes of food go to waste annually across the world, plastic packaging is proving an ideal solution in helping to address this waste. In addition, plastic's resistance to moisture, cold or heat facilitates convenient storage and distribution. Some uses of plastic packaging in the food industry are shown in Fig. 2.3.

Furthermore, plastic packaging enables easy labelling and conveying of information to potential consumers about the product, such as details about the use, safety and nutritional value of the food item.

2.1.3 Plastic is protecting lives

Plastic is applied in the health sector, where it has made life and the sector safer and cost-efficient. Syringes, insulin pens, surgical gloves, intravenous (IV) tubing, among many other things, are medical supplies and products made from plastic. Its availability helps to reduce the need to sterilise and reuse equipment, which aids in limiting the transmission of infection. Also, plastic helps to increase hygiene at hospitals and medical facilities. These facilities are maintained, secured and protected from infections with vinyl-based wall coverings, ceilings and flooring through the special qualities of the polymer that improve cleanliness.

Plastic improves safety. One of its properties is strength, and so it can be used in making tamper-proof caps for waste bags, blister packs and medical packaging. It is not easily broken, making it a dependable material for storage and transit. Plastic provides comfort and flexibility, as medical components made with alternative materials, such as metal, are not flexible and usually cause discomfort, for example, in prosthetics. Furthermore, medical components made with plastic are more cost-efficient as these components can be mass-produced due to the relatively low cost of plastic. Medical products made of plastic are reliable as they are resistant to wear, tear or corrosion. Fig. 2.4 shows some applications of plastic in healthcare.

Plastic has contributed immensely to healthcare, particularly due to its ability to protect. Plastic personal protective equipment (PPE), which includes gloves, respirators, goggles, masks, gowns and many other items, plays a fundamental role in providing protection for health personnel in their work,

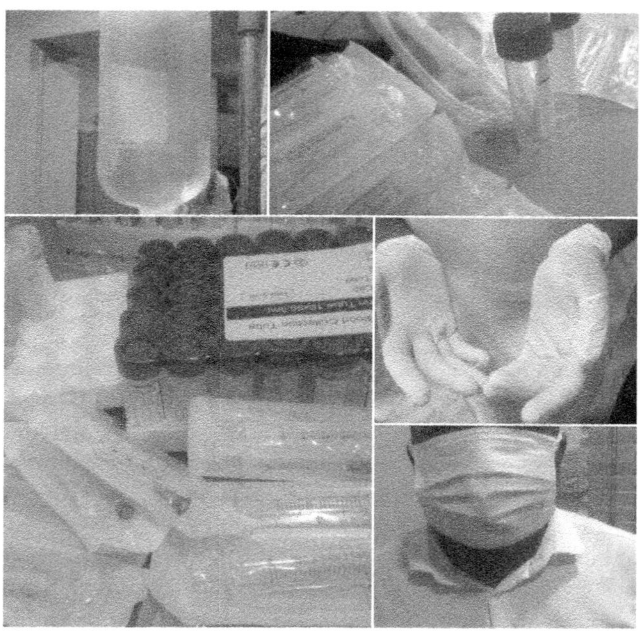

FIGURE 2.4 Plastic in healthcare.
Plastic application in healthcare. Photo credit: Kunle Olatayo.

and likewise the public when in contaminated spaces. This was evident during the recent COVID-19 pandemic. Plastic is a critical material used for PPE production, which contributed significantly to reducing the impact of the pandemic. During the pandemic, PPE was made mandatory for health workers and other personnel on the frontline for protection and in handling patients. Countries across the globe made it mandatory for the public to protect themselves with face masks while in public places. This led to the daily production and consumption of PPE in millions globally for protection against the pandemic virus. Approximately 89 million medical masks, 76 million gloves and 1.6 million goggles were required monthly for health workers during the period (World Health Organization, 2020).

2.1.4 Plastic: a double-edged sword

While plastics have been described as having several benefits considering their unique properties, the unsustainable continued increase in the production and use of plastics is inadvertently producing a corresponding increase in the amount of plastic waste generated. High plastic production can be linked to high resource consumption and subsequent high waste generation. As more diverse plastic polymers, with several various additives, flow into the economy, the environmental impact and burden borne by society through the consumption of plastic products increase. The complete mismanagement of this non-biodegradable plastic results in the resource being detrimental to the environment.

The ever-increasing mass of plastic waste ending up in food samples, in the land and in water is argued to have significant negative environmental impacts on ecosystems, biodiversity and health. This is because the degradation of plastic waste is difficult as a result of its polymer structure (Chen et al., 2019). Likewise, the durability of polymers adds to the burden of products improperly disposed of in the environment.

Unfortunately, the ensuing plastic waste from several unsustainable activities amounts to a considerable percentage of the total global waste stream. The proliferation in the use of plastics, together with the mismanagement of the end-of-life of plastics, has resulted in prevalent plastic pollution. The ongoing increase in plastic waste being generated and the resulting landfilled waste and pollution continue to be of great concern to governments, stakeholders and the public across countries. Most plastics produced and consumed end up either in landfills or incinerators for energy recovery, or are recycled at best, albeit in a limited quantity. Approximately 8.3 billion tonnes of plastic were produced globally between 1950 and 2015, which resulted in an estimated generated plastic waste of 6.3 billion tonnes, about 79% of which ended up in landfills and open dumps (Geyer et al., 2017).

Currently, plastic products are being designed to be reused or recycled. This helps reduce the environmental burden of plastics while retaining their usefulness and value, and this calls for a shift from the linear economy of 'produce, use and discard' to a more sustainable and circular model for plastics. The linear model represents not only a significant loss of resources but also continued destruction to the environment. However, the circular economy provides a promising opportunity where plastic waste is diverted from the landfill to a better and more sustainable plastic waste economy.

2.2 Plastics across the value chain

From the extraction of raw materials for plastics production to eventual plastic waste management, each stage of the plastics value chain plays an essential role. This chain is complex, impacting most, if not all, industrial sectors. Understanding the value chain is critical for making informed decisions about reducing waste, optimising resource use, implementing circular economy practices and addressing environmental impacts. The representation of the plastics value chain in Fig. 2.5 demonstrates the networks and sequence of the different value chain stages, as well as the flow of plastic waste resources across the value chain.

The stages of the plastic value chain are plastic material production, plastic product manufacturing, plastic product use and waste generation, plastic waste collection and handling, plastic waste sorting, plastic waste recycling and plastic recyclate market (Olatayo et al., 2023).

2.2.1 Plastic material production

This is the first stage of the value chain, and it represents all the practices and processes leading to the production of primary plastics, otherwise known as polymers or resins. Plastic is produced mainly from petrochemicals, derived from fossil fuels. The process starts with the raw material extraction, followed by a separation process of converting components of chemicals into olefins (The Plastics Landscape: Risks and Opportunities Along the Value Chain, 2019). However, a very small share of production is bio-based plastics. As illustrated in

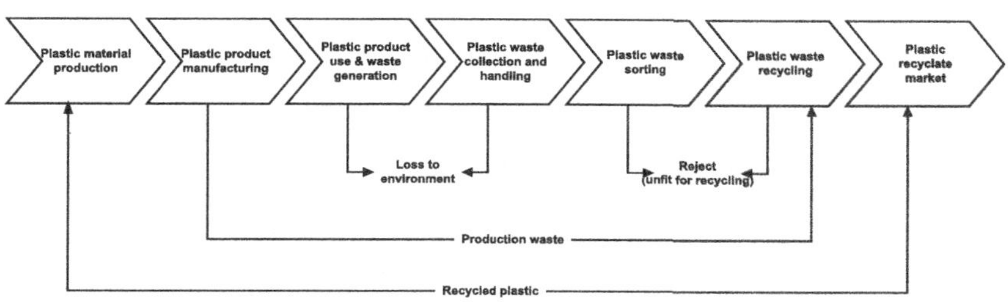

FIGURE 2.5 Plastic value chain.
Diagram of the plastic value chain.

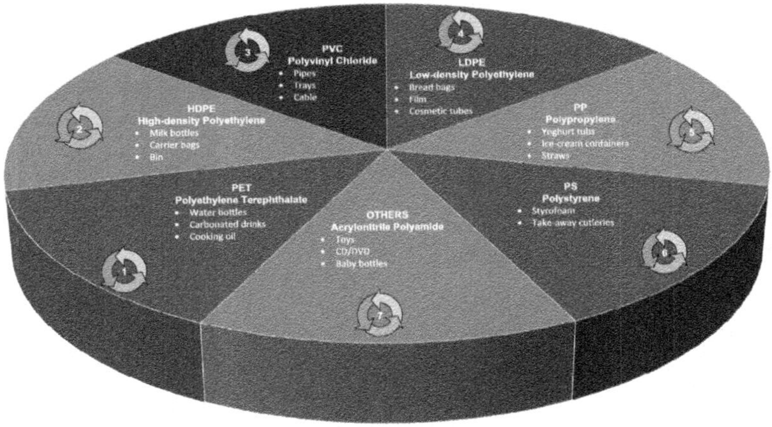

FIGURE 2.6 Different types of primary plastics.
Plastic types.

Fig. 2.6, primary plastics include polyethylene terephthalate (PET), high-density polyethylene (HDPE), polyvinyl alcohol (PVC), low-density polyethylene (LDPE), polypropylene (PP), polystyrene (PS) and others (acrylonitrile–butadiene–styrene (ABS), styrene acrylonitrile (SAN), polyamide (PA), etc.).

2.2.2 Plastic product manufacturing

The plastic product manufacturing stage comprises several production steps that convert primary plastics into various plastic products through processes such as injection moulding, extrusion, blow moulding and thermoforming, to form plastics into consumer goods, packaging materials, electronics, automotive components and several other products. Processes at this stage start with the transformation of the resin raw material into plastic products and residue. Resins are obtained from either petrochemicals or plastic recycling to manufacture plastic and plastic-containing products. Product design is an important aspect of this stage of the value chain. During design, the qualities and functions of the plastic products are determined, such as polymer type, recyclability, shape and colour. Interestingly, the level of production of plastic products has increased significantly over the previous 50 years, caused by a global shift to single-use plastic packaging from reusable containers, and stimulated partly by the rise in convenience purchasing and takeaway consumption.

2.2.3 Plastic product use and waste generation

This stage describes the several sectors of the industry that use or consume plastics. Numerous sectors make use of plastics, including construction, electrical and electronic, health and healthcare, packaging, automotive and furniture. Four of these sectors, namely, packaging, electrical and electronic, automotive and construction, account for three-quarters of plastics consumed globally (The Plastics Landscape: Risks and Opportunities Along the Value Chain, 2019).

2.2.4 Plastic waste – waste collection and handling

The extremely high production and consumption of plastics is associated with the huge generation of plastic waste. The ongoing increase in the use of plastics contributes significantly to the total volume of waste being generated within the environment. A substantial percentage of plastics usually goes to waste, with the main proportion of this either landfilled, incinerated, or leaked into the environment. Regrettably, the ever-increasing quantity of plastic waste being generated enters the land, water bodies and food samples, with negative environmental impacts on the ecosystems, human health and biodiversity.

Considering that single-use packaging and other short-lifespan products constitute a substantial percentage of plastic production annually, these products become waste within the production year. Furthermore, considerable volumes of plastic waste are usually imported from countries where they are not treated locally. The ban on plastic waste import from mostly developed economies by China has resulted in the diversion of this waste to developing countries, mainly in Southeast Asia and some sub-Saharan countries such as Senegal and Ethiopia.

Plastic waste collection and handling is an essential and core component of the plastic waste management system universally. This stage is crucial in overall plastic waste recycling and management. The collection of plastic waste depends on regional differences, supported by infrastructure that differs considerably among countries and regions. Most countries in the Global South apply plastic waste management supported by a still-developing and informal collection system, unlike those in the Global North with advanced infrastructure and formal collection systems. However, countries usually implement different collection schemes to suit their circumstances and environments.

In the Global North, the collection systems include kerbside and bring-to sites. Kerbside, also known as door-to-door, is the most commonly used. In this collection system, households are provided with bins or bags as waste receptacles, which are placed at the door or kerb on collection days, and this can be for mono or mixed plastic streams. The bring-to site involves the accumulation of waste at central or communal collection sites with a bigger capacity. In many developing economies, several methods of waste collection are practised, which include kerbside, communal, drop-off and buy-back, with the informal sector playing a major role in waste collection.

2.2.5 Sorting

Sorting is a crucial aspect of plastic waste recycling, as it has an impact on the quality of recycling, which is the next stage of the value chain. It is a process that can be conducted manually or automatically, using advanced imaging-informed sorting facilities, for example, near infra-red (NIR). Without advanced sorting technology, there is less efficiency in separation, which results in target plastic waste being rejected at about 13% (Hahladakis et al., 2018).

The challenge of sorting depends greatly on the composition of the plastic waste. Mixed plastic waste streams could include flexible plastics, multi-layered plastics and bio-based plastics, which are mostly non-NIR identifiable. Furthermore, sleeves and labels usually make up over 60% of the components of plastics (Hahladakis and Iacovidou, 2019). This can result in faults where the sorting equipment finds it difficult to identify types of polymers. A difference in the polymer type between the sleeve and the component can lead to rejection. Equally, a similarity in the polymer type between the sleeve, particularly a coloured sleeve, and the component can decrease the quality. Another challenge at the sorting stage could be different materials being mixed with plastic waste.

2.2.6 Plastic recycling

Plastic waste is recycled after the sorting process. The recycling of plastic waste involves two main processes, namely mechanical and chemical. The mechanical process of recycling entails the breakdown of the physical state of the plastic waste by shredding while retaining the chemical state. Recycling through the chemical process has to do with the breaking down of the plastic waste chemical forms by melting using gasification, pyrolysis, or depolymerisation.

Specific factors determine the degree of quality of the recyclate and the ease of recycling, notwithstanding the different processes for recycling plastic waste. These factors include plastic type, design, properties, colour and mixed polymer issues. When different non-compatible plastic polymers are mixed, the property difference of these polymers becomes a major challenge in the production of plastic recyclates from waste.

The existence of challenges in the plastic recycling stage of the value chain can result in down-cycling, in which recyclate quality is lower than that of the original plastic material undergoing recycling. Also, instances are found where certain plastic materials are not suitable for multiple or repeated recycling. As a result, recyclate quality is a significant determinant when it comes to the substitution of virgin resins with recycled ones in plastic product production and consumption. The plastic recycling sector can be regarded as relatively struggling and fragmented in comparison to the recycling of other materials such as glass, paper and metal (Ranta et al., 2018; Milios et al., 2018).

2.2.7 Recyclate market

The development of a reliable and transparent market where plastic recyclate can be sold is crucial. This comes with having a high-value recycling system where plastic recyclates produced are of high purity in terms of the plastic type and colour. If contamination affects the production and quality declines, the recyclate becomes worthless and unsellable in the market. Increasing data and information on laboratory test results for recyclate and quality attributes could promote more transparency and trust. This data and information can include recyclate purity, technical specifications, colour, additives, qualitative certificates and availability of the recyclate supply in terms of quantity and time.

2.3 Plastics in the circular economy

The circular economy is an economic model that promotes the shift from the traditional linear model of resource consumption based on 'take-make-use-dispose' to restoration and regeneration (Ellen MacArthur Foundation, 2015). It removes or reduces the constraints of 'end-of-life' associated with the linear economy during production, use and disposal. The linear economy model largely results in millions of tonnes of plastic waste ending up in landfills, open dumps and water bodies, or being burnt in some instances. There is a necessity to rethink the way plastics are made, used and discarded. Essentially, this requires redesigning a plastics value chain system that functions in practice, with minimum plastic waste and loss of environmental and economic value.

The circular economy addresses and restores the end-of-life of a material with the principle of reduce, reuse, recycle and recover, within the processes of production, distribution and consumption (Kirchherr et al., 2017). The model supports and aligns with sustainable development and the circulation of materials or products in closing the loops (Blum et al., 2020), thereby prescribing the review of resource production, consumption and recovery.

The circular economy, a concept centred on the principles of the waste hierarchy, is increasingly being adopted across countries universally as a major dynamic application to reduce the quantity and impact of plastic waste in the environment. The main idea behind the concept is the principle of cradle-to-cradle, which focuses on closing the biological and technical loops of materials, plastics inclusive.

The application of the circular economy principle to plastics creates an economy where plastic waste is controlled or eliminated by preventing the material from becoming waste, or it is reused or recycled (Iacovidou et al., 2017). As opposed to the traditional plastic economy, the circular plastic economy promotes sustainable plastic product design and manufacture, intensive use, reuse, repair, refurbishment and recycling, thereby advancing a sustainable and eco-effective system that prevents or significantly reduces plastic from being degraded into waste. This ensures that pathways where plastic manufacturing, use, reuse and recycling practices and processes that eliminate or considerably limit plastic waste generation are identified, adopted and implemented.

The circular plastic economy is immensely beneficial to the environment as the quantity or frequency of extraction and use of plastic resources for production is reduced, and consequently, the mass of plastic waste generated and disposed of into the ecosystem is minimised. Pressures on plastic raw materials for production decrease as a result of the reduction in extraction, and energy savings are equally recorded. The circularity of plastics also contributes to the mitigation of environmental plastic pollution and limits the emissions of poisonous gases and substances such as carbon monoxide, acid gases and heavy metals into the atmosphere.

2.4 Potential for circularity across the plastics value chain

The vision for a circular plastic economy will involve the adoption and integration of circularity across the complete plastics value chain system, including the different value chain stages. This is because the contributing performances of the respective stages of the plastics value chain are central to the growth of the system and the advancement of the circular plastic economy (Lüdeke-Freund et al., 2019). Each stage in the plastics value chain has implications for any process or outcome of the system, and this must be considered when assessing the system. Therefore the complete development or circularity of the plastics value chain system cannot be achieved by a singular stage, but rather all the different stages of the chain and their associated practices and activities must be included (Paletta et al., 2019).

Enabling circular economy practices across the plastics value chain involves assessing and identifying opportunities and challenges existing in every stage of the value chain for the implementation of circularity across the system. The idea is to have a value chain where primary plastics and plastic products are used efficiently, waste is reduced and value is added to the system. The aspects of the plastics value chain analysed include production, use and end-of-life.

2.4.1 Circularity in production

Circularity can be integrated at the production stage of plastics, particularly in the design of the plastic product. Plastic products can be designed with less plastic material, or plastic recyclates or recycled plastic can be incorporated as part of the design, or plastic products can be designed so that they can be recycled after end-of-life. However, these different options have their associated challenges. For instance food packaging must meet a high standard, including the ability to keep packaged food fresh and toxin migration low, and it must be lightweight, which may be difficult when using less plastic material or using plastic recyclate in the design (Masmoudi et al., 2020). Plastic recyclate might be challenging to use in making new plastic products due to the possibility of contamination of polymer types, affecting toxicity, durability and weight (Foschi et al., 2020). Mixed polymers and contamination are among the most challenging difficulties when incorporating plastic recyclates into new plastic products. Plastic recyclates are usually made of different types of polymer and contaminated with additives and other non-plastic materials, thereby resulting in a number of complications, such as lumps; this limits the value of the final product.

Some of these difficulties can be overcome by integrating the overall plastics value chain in new plastic product designs in order to facilitate circular and improved production, use and waste recyclability. A number of principles are required regarding the design of plastic products with the intention to phase out waste and ensuring the recyclability of products. These include designing long-life plastic products and product life extension to slow resource loops; using plastic waste recycling to close resource loops; designing with consideration for sustainable sourcing; plastic product design centred on optimal resource use, environmental friendliness, safe use, prolonged product use and recycling.

2.4.2 Circularity in plastic use

This aspect of the value chain addresses the demand, consumption and generation of plastic waste during the use phase. A major challenge to the integration of circularity here is the wide gap between the consumers' understanding and perception of the environmental implications of the end-of-life of plastic products and the actual impact on the environment. Consumers need to be more aware and better informed of the implications of their consumption and the negative impacts of their choices on the environment in order to overcome this challenge. Many consumers are concerned about the environmental impact of plastics, demonstrating general awareness among the populace regarding plastic pollution. Habits, knowledge, beliefs and social norms are what usually influence the consumers' demand for and consumption of plastic products. Furthermore, consumers are at the forefront when it comes to the design of new plastic products and an increase in plastic recycling rates. Therefore considering the role of consumers in initiating demand for plastic products and separating plastic waste, behavioural change is critical for the successful integration of circularity in the value chain, and this must be taken into account when formulating circularity solutions for the plastics value chain.

2.4.3 Circularity at end-of-life

Plastic circularity cannot be thoroughly examined without assessing end-of-life solutions for plastic waste. The circular economy at the end-of-life phase covers the collection, sorting and recycling of plastic waste. A very important aspect of circularity in the value chain is the looping of the plastic waste back into the production of new plastic products.

Collection and sorting are the first phases of plastic waste management. Circularity can be integrated through separate waste collection, tracer-based sorting and deposit refund schemes, which will improve plastic waste recycling significantly. Collection systems should be promoted in which all plastic waste is separated in standardised waste bins for different types of waste and then recycled. This is contrary to the system currently used in developing countries, namely, a common single bin for all types of waste material. A multi-bin kerbside collection system that requires plastic waste to be taken to material recovery facilities for sorting before recycling or incineration, as applied in developed countries, should be encouraged.

For example, the one-bin-to-rule-them-all system in the United Kingdom could increase the plastic waste collection rate and decrease the leakage of waste into the environment, using circularity-centred infrastructure for product design, recycling and sorting (Burgess et al., 2021). Separate waste collection of plastic waste would enable improved and results-driven communication with consumers, which would make managing plastic waste easier. The application of tracer-based sorting of plastic waste will enable easy sorting of the waste according to different polymer types, thereby reducing contamination that would affect recycling.

Deposit refund schemes are another strategy that can bring circularity to plastic waste collection. The introduction of such a system, for example, PET bottles, would result in a 'cleaner' plastic material stream, as only distinct plastic products would be accepted, thereby advancing circularity. However, this type of scheme requires proper management, as managing a large array of plastics would be challenging and demanding. Also, the scheme can be complex to set up and is not suitable for all plastic products.

Furthermore, recycling is a major aspect of the value chain that greatly contributes to a circular economy. The recycling of plastic waste has been promoted for integration and advancement of the circular economy, which aims for the conversion of plastic waste into new material and its reintroduction as a new product into the system (Hahladakis and Iacovidou, 2019). This solution for managing plastic waste is important in tackling plastic pollution in the environment and atmospheric emissions. The types and processes of recycling have an impact on the quality of recycled plastic or recyclate. Mechanical recycling is the major recycling technique used in advancing circularity, with chemical recycling usually being applied where plastic waste, such as composite materials, cannot be mechanically recycled. Both recycling technologies help to increase total plastic recycling rates. Biological recycling has potential but requires more development to make an impactful contribution to plastic recycling and the circular economy.

2.5 Chapter summary

In this chapter, an overview was presented of the contribution of plastics to the economy and the vision of a circular economy. The role and contributions of plastic material in the economy, the plastics value chain, the different stages in the value chain, plastics in the context of the circular economy and the various opportunities and challenges of the system with respect to the integration of circularity across the value chain were discussed.

Plastic has extensive use and several roles in products and applications. The advancement in its properties has made it more applicable in diverse sectors of the economy, including automotive, health, textile, agriculture, furniture, construction and electronics. However, its life cycle, which covers production, use and end-of-life, can be linked to several problems, such as greenhouse gas emissions during production, unsustainable resource use and environmental pollution from plastic waste. The increased production and use of plastic, combined with poor management of the corresponding waste, have resulted in the accumulation of plastics in landfills and in our water systems, with the associated negative impacts on human health, biodiversity and the environment.

The circular economy model is experiencing increasing focus globally to reduce the quantity and impact of plastic waste in the environment. It supports and aligns with the sustainable development and circulation of materials or products with the idea of closing material and value loops. The application of the circular economy principle with regard to plastics creates an economy where plastic waste is controlled or eliminated by preventing the material from becoming waste, or it is reused or recycled. As opposed to the linear plastic economy, the circular plastic economy promotes sustainable plastic product design and manufacture, intensive use, reuse, repair, refurbishment and recycling. It is extremely beneficial to the environment as the quantity or frequency of extraction and use of plastic resources for production is reduced, and the quantity of plastic waste generated and disposed of decreases.

The circular economy needs to be integrated into the plastics value chain. This requires identifying opportunities and challenges at all stages of the value chain and then implementing circularity. These value chain stages are plastic material production, plastic product manufacturing, plastic product use and waste generation, plastic waste collection and handling, plastic waste sorting, plastic waste recycling and the plastic recyclate market. From the extraction of raw materials for plastics production to eventual plastic waste management, all the value chain stages play an essential role. Therefore understanding the plastics value chain is critical for informed decisions regarding reducing waste, optimising resource use, implementing circular economy practices and addressing environmental impacts.

A core aspect of circularity in the value chain is the looping of the plastic materials back into the production of new plastic products, and reusing plastic materials and products.

References

Blum, N.U., Haupt, M., Bening, C.R., 2020. Circular doesn't always mean "Sustainable" Resources, Conservation and Recycling 162, 1–10. Available at https://doi.org/10.1016/j.resconrec.2020.105042.

Bucknall, D.G., 2020. Plastics as a materials system in a circular economy. Philosophical Transactions of the Royal Society A: Mathematical, Physical and Engineering Sciences 378 (2176), 20190268. Available at https://doi.org/10.1098/rsta.2019.0268.

Burgess, M., Holmes, H., Sharmina, M., Shaver, M.P., 2021. The future of UK plastics recycling: one bin to rule them all. Resources, Conservation and Recycling 164. Available at https://doi.org/10.1016/j.resconrec.2020.105191.

Chen, Y., Cui, Z., Cui, X., Liu, W., Wang, X., Li, X.X., et al., 2019. Life cycle assessment of end-of-life treatments of waste plastics in China. Resources, Conservation and Recycling 146, 348–357. Available at https://doi.org/10.1016/j.resconrec.2019.03.011.

Ellen MacArthur Foundation, 2015. Growth within: A circular economy vision for a competitive Europe. McKinsey Center for Business and Environment. Available at https://content.ellenmacarthurfoundation.org/m/3d4eba36b0311c08/original/Growth-within-A-circular-economy-vision-for-a-competitive-europe.pdf.

Foschi, E., Zanni, S., Bonoli, A., 2020. Combining eco-design and LCA as decision-making process to prevent plastics in packaging application. Sustainability (Switzerland) 12 (22), 1–13. Available at https://doi.org/10.3390/su12229738.

Geyer, R., Jambeck, J.R., Law, K.L., 2017. Production, use, and fate of all plastics ever made. American Science Advances 3 (7), e1700782. Available at https://doi.org/10.1126/sciadv.1700782.

Hahladakis, J.N., Iacovidou, E., 2019. An overview of the challenges and trade-offs in closing the loop of post-consumer plastic waste (PCPW): focus on recycling. Journal of Hazardous Materials 380, 120887. Available at https://doi.org/10.1016/j.jhazmat.2019.120887.

Hahladakis, J.N., Velis, C.A., Weber, R., Iacovidou, E., Purnell, P., 2018. An overview of chemical additives present in plastics: migration, release, fate and environmental impact during their use, disposal and recycling. Journal of Hazardous Materials 344, 179–199. Available at https://doi.org/10.1016/j.jhazmat.2017.10.014.

Iacovidou, E., Millward-Hopkins, J., Busch, J., Purnell, P., Velis, C.A., Hahladakis, J.N., et al., 2017. A pathway to circular economy: developing a conceptual framework for complex value assessment of resources recovered from waste. Journal of Cleaner Production 168, 1279–1288. Available at https://doi.org/10.1016/j.jclepro.2017.09.002.

Kirchherr, J., Reike, D., Hekkert, M., 2017. Conceptualizing the circular economy: an analysis of 114 definitions. Resources, Conservation and Recycling 127, 221–232. Available at https://doi.org/10.1016/j.resconrec.2017.09.005, https://doi.org/10.1016/j.resconrec.2017.09.005.

Lüdeke-Freund, F., Gold, S., Bocken, N.M.P., 2019. A review and typology of circular economy business model patterns. Journal of Industrial Ecology 23 (1), 36–61. Available at https://doi.org/10.1111/jiec.12763.

Masmoudi, F., Alix, S., Buet, S., Mehri, A., Bessadok, A., Jaziri, M., et al., 2020. Design and characterization of a new food packaging material by recycling blends virgin and recovered polyethylene terephthalate. Polymer Engineering & Science 60 (2), 250–256. Available at https://doi.org/10.1002/pen.25278.

Milios, L., Holm Christensen, L., McKinnon, D., Christensen, C., Rasch, M.K., Hallstrøm Eriksen, M., 2018. Plastic recycling in the Nordics: A value chain market analysis. Waste Management 76, 180–189. Available at https://doi.org/10.1016/j.wasman.2018.03.034.

Olatayo, K.I., Mativenga, P.T., Marnewick, A.L., 2023. Plastic value chain and performance metric framework for optimal recycling. Journal of Industrial Ecology 27 (2), 601–623. Available at https://doi.org/10.1111/jiec.13384, https://doi.org/10.1111/jiec.13384.

Paletta, A., Leal Filho, W., Balogun, A.-L., Foschi, E., Bonoli, A., 2019. Barriers and challenges to plastics valorisation in the context of a circular economy: case studies from Italy. Journal of Cleaner Production 241, 118149. Available at https://doi.org/10.1016/j.jclepro.2019.118149.

Ranta, V., Aarikka-Stenroos, L., Ritala, P., Mäkinen, S.J., 2018. Exploring institutional drivers and barriers of the circular economy: a cross-regional comparison of China, the US, and Europe. Resources, Conservation and Recycling 135, 70–82. Available at https://doi.org/10.1016/j.resconrec.2017.08.017.

The plastics landscape: risks and opportunities alon g the value chain An investor initiative in partnership with UNEP Finance Initiative and UN Global Compact. (2019).

World Economic Forum, 2016. The New Plastics Economy: Rethinking the Future of Plastics. World Economic Forum. Available at https://www3.weforum.org/docs/WEF_The_New_Plastics_Economy.pdf.

World Health Organization, 2020. Shortage of Personal Protective Equipment Endangering Health Workers Worldwide. World Health Organization. Available at https://www.who.int/news/item/03-03-2020-shortage-of-personal-protective-equipment-endangering-health-workers-worldwide.

Chapter 3

Plastic waste and the waste management hierarchy

3.1 Managing plastic waste

The production and use of plastic, combined with the poor management of the corresponding waste, have contributed to the accumulation of waste in landfills, widespread pollution and other negative impacts on human health, biodiversity and the environment. The massive volume of plastic waste being generated from the consumption of a broad range of plastic products has to be managed properly to prevent plastic pollution and promote sustainability of the environment.

The management of plastic waste is complicated by challenges in setting up effective systems for waste collection, separation at source, sorting and recycling of mixed waste. This results in the largest share of plastic waste generated ending up in landfills and open dumps. In landfills, plastic waste accumulates for a long time since it is non-biodegradable. Thus plastic pollution has become an environmental challenge globally. About 70% of the global plastic waste was uncollected in 2023, and this was either disposed of in landfills, subjected to open burning or leaked into the environment (Berger, 2023). The continuous accumulation of plastic waste in landfills and illegal open dumps is a threat to the ecosystem. While the reduction of plastic waste generation is essential, the efforts and the drive to manage existing waste are critical in addressing plastic pollution.

Plastic waste management can be described as the process of handling and managing the waste generated from the consumption of plastics, with the purpose of reducing the volume generated, decreasing pollution, reducing its impact on the environment and promoting sustainability. It includes the collection, handling, transportation, processing and eventual disposal of plastic waste. In improving waste management systems and moving towards a circular plastic economy, legislation and measures are implemented by governments, local authorities, communities, interest groups and other relevant stakeholders.

Considering the prevalence of plastic waste in the environment as well as the presence of various challenges driving its mismanagement, intervention is required to tackle these challenges and increase the circularity of plastic. This

Plastic Circularity. DOI: https://doi.org/10.1016/B978-0-443-40667-6.00002-0

will consequently limit the burden or negative impact of plastic on the natural environment, biodiversity and human health. The intervention requires a rethink of how plastics are being used and the corresponding waste is being managed.

The foremost aspect of this intervention in managing plastic waste should incorporate fundamentals that involve the avoidance or prevention of plastic waste from occurring in the first instance, or the use of alternative materials that are environmentally friendly as a substitute for plastics. Also, where possible, opportunities or measures should be promoted that would cut the unnecessary or excessive consumption of plastics. Where it is not possible, an increase in the manufacturing efficiency of plastic products to reduce the material content should be considered and advanced.

3.2 Waste management hierarchy systems

In the search for a friendlier and more sustainable environment, the waste hierarchy represents a guiding principle and roadmap to effective waste management and the sustainability of the environment. Specifically, the circularity and sustainability of plastics can be achieved by implementing waste management as defined in the waste management hierarchy system. The system promotes a circular economy where plastics are designed innovatively and sustainably for maximum use, reuse, repair, refurbishment, remanufacturing, repurposing and recycling.

The application of the waste management hierarchy provides benefits which include the conservation of resources and resource efficiency, as well as reduction in energy consumption, plastic waste and pollution and impact on the environment. Interestingly, the waste management hierarchy system has evolved and become broader over time. The background, types, hierarchy options and significance of the waste hierarchy frameworks are explored further in subsequent sections.

3.2.1 European Union waste hierarchy

The environmental and waste policies of the European Union (EU) are implemented through the European waste hierarchy. This hierarchy articulates principles for reducing environmental damage from improper and unsustainable waste management and outlines different levels of best to worst options for waste management. With prevention being the most desirable option, at the top of the hierarchy, and disposal being the least desired, at the bottom of the hierarchy, these varied options define a priority order for practices and processes of waste management.

The waste hierarchy is a practical implementation of the principles of industrial environmental management and has altered waste economies dramatically. It aims to provide a status change for waste, with respect to product design, manufacturing and consumption, using the principles associated with industrial ecology. This hierarchy has characterised a significant change in the practices and activities of waste management in Europe, considering that about 70% of waste ended up in landfills in the late 20th century (Giusti, 2009). The hierarchy can be described as normative, considering it ranks how desirable waste management practices are; it can also be described as dynamic, as it may be used to determine which combination of options provides the most complete benefits or advantages in terms of the environment, economy and society. It enables the coordination of actions across organisational and institutional boundaries, as well as the vertical integration of the various phases of the life cycle of a product.

The Waste Framework Directive of the European Union contributed to the waste legislation in Europe through the establishment of the waste hierarchy. The first iteration of the waste hierarchy dates back to the Council Directive on Waste of 1975 (75/442/EEC) (Council Directive of 15 July 1975 on waste 75/442/EEC, 1975). It classified two waste management methods, without providing a hierarchy or priority order of options: (1) reducing waste amounts; and (2) disposing of waste through reuse, recycling, and storage and underground. The first hierarchy for the management of waste was developed in 1979, where the concept of Ladder van Lansink (Ladder of Lansink) was proposed by Ad Lansink (Pires and Martinho, 2019; Zhang et al., 2022). This evolved into the current waste hierarchy, with the Waste Framework Directive updated (91/156/EEC) in 1991. In 2008 the Waste Framework Directive 2008/98/EC introduced the universally recognised waste hierarchy. The waste hierarchy was later inverted into the national laws of the Member States of the EU (Committee of the Regions, 2015). Fig. 3.1 illustrates how the EU waste hierarchy evolved over time.

The 2008/98/EC directive covers different management measures or options that could be applied to a material or product before and after it

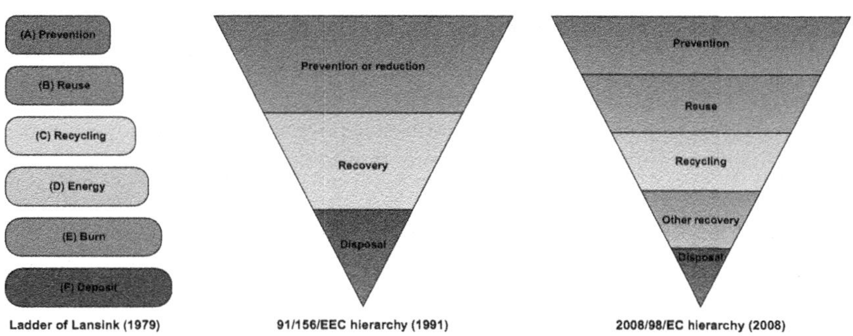

FIGURE 3.1 The evolution of the EU waste hierarchy. The changes that happened during the lifecycle of the EU waste hierarchy.

becomes waste. Article (4) of the directive defines the priority order of the options of the waste hierarchy to include five measures or options: prevention (most desirable option), preparing for reuse, recycling, other recovery (e.g. energy recovery) and disposal (least desirable option).

The waste hierarchy supports the transition of the EU to a circular economy, with a progressive decrease in the mass of waste going to landfill. By following the top-down priority guidelines, countries are expected to design and implement plans and policies to promote and enable a shift towards the options higher up the waste hierarchy. More recently, the Waste Framework Directive 2008/98/EC has been amended by the Waste Framework Directive 2018/851, which strengthens the requirements for the prevention of waste substantially.

The 2008/98/EC provides definitions for the different options in the waste hierarchy. In addition, more explanation of the options and related terms is provided in a related document, 'Guidance on the interpretation of key provisions of Directive 2008/98/EC on waste', issued in 2012 by the Director General of Environment of the European Commission.

Prevention is the option rated at the top of the waste hierarchy as being the most preferable. The Waste Framework Directive 2008/98/EC describes it as measures or actions undertaken to 'reduce' the amount of waste, and these actions include 'reuse and the increase of the life span of products'. Similarly, the Department for Environment, Food and Rural Affairs (DEFRA) describes it as comprising three clear actions: 'replacement' of hazardous materials with materials that are less hazardous, 'reduction' using less material and 'reuse'. The Waste and Resources Action Programme (WRAP) describes it as a measure involving the three actions of 'avoidance', 'reduction' and 'reuse'. The prevention of waste involves practices such as design, production and consumption that will not result in the generation of waste. This can be the design for reuse and recycling, avoiding spill or waste in production and actions by consumers to promote repair, maintenance, buy-back and second-hand retailing.

Next, in descending order of priority in the waste hierarchy, is preparing for reuse. It is described by DEFRA as actions that involve 'checking', 'repairing', 'cleaning', 'repairing' and 'refurbishing' spare parts or whole products. WRAP further clarifies that checking and cleaning are attributes of preparing for reuse, and repairing or refurbishing are actions that can lead to the reuse of waste products.

Recycling is the third level. The Waste Framework Directive 2008/98/EC describes it as a measure of 'reprocessing' waste materials into products, substances or materials for either the original or other purposes. Several other descriptions have further termed it as up-cycling or down-cycling. Up-cycling has to do with reprocessing waste into products, items or materials of higher value or purpose than the original, whereas down-cycling involves reprocessing waste into products, items or materials with lower value or purpose than the original.

Other recovery follows in the waste hierarchy. This option is not defined in the Waste Framework Directive 2008/98/EC; instead, an example, 'energy recovery', is given. WRAP and DEFRA term it as 'recovery of materials' and 'energy recovery' from waste, respectively.

Disposal is at the bottom of the waste hierarchy, being the least preferable option. The description is related to a landfill.

3.2.2 R-framework hierarchy

The R-framework for waste management has been analysed and related to the waste hierarchy system, both in theory and in practice. More importantly, it has been strongly termed as the central principle and guideline of the circular economy. There are different R-frameworks and they function on the principle of the waste hierarchy, with a descending order of priority, of the application of the varied 'R' options. The different R-frameworks are similar and related to the waste hierarchy, yet they differ in the number of circularity approaches or strategies they offer. Notable R-frameworks promoted in the literature include both the 3R (Ghisellini et al., 2016) and 6R (Yan and Feng, 2014; Jayal et al., 2010). Some studies have further explored the 9R framework (Van Buren et al., 2016; Potting et al., 2017), which is characterised by 10 waste management options, with the first and last options regarded as R0 and R9, respectively. It is an offshoot of the 6R framework with the options of refuse, rethink, reduce, reuse, repair, refurbish, remanufacture, repurpose, recycle and recover, in descending order of preference or priority. The 3R and 6R frameworks are illustrated in Fig. 3.2.

3.2.2.1 The 3R hierarchy

The 3R waste hierarchy consists of three options for waste management practices, namely reduce, reuse and recycle, in descending order of priority, as represented in Fig. 3.2. This signifies their order of importance. Firstly, waste is expected to be reduced – this is the most preferred option – followed by reuse and recycle as the second and third options in that order of hierarchy. The 3R seeks to produce the least quantity of waste while maximising the useful advantages of products. The principle works to minimise the negative consequences of waste by taking actions to reduce the amount of waste, to reuse waste materials or products with simple means of treatment and to recycle waste where it is used as resources to create new or transformed products. The long-term application and integration of these varied options will address waste minimisation effectively and sustainably.

Reduce, which is the first option, can be achieved by avoiding or limiting the use of raw materials. It involves all practices targeted at decreasing the

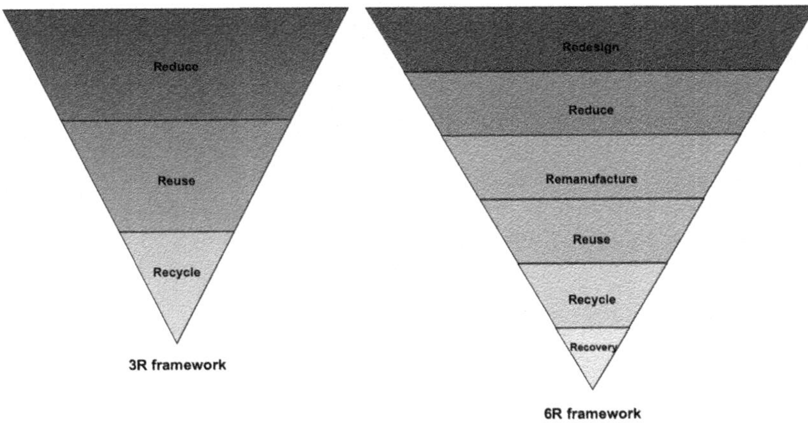

FIGURE 3.2 **The 3R and 6R framework waste hierarchies.** Comparison of 3R and 6R waste hierarchies.

quantity of production waste. It emphasises the judicious application of resources in production. Also, campaigns to inform, enlighten and motivate the public to use products that generate less waste, such as the use of bio-bags instead of plastic bags, can be promoted. Reduction can be realised by introducing legislation, local schemes and programmes and sustainable product designs to keep compostables and recyclables away from waste. Legislation compelling manufacturers to apply specific standards in the design of products or limiting production capacity has been adopted in the United States and Europe (Ahmadi, 2017).

The second option of reuse takes effect where reduction is not feasible. It does not involve reprocessing. This option could involve the repair, sales or donation of used products to community or charity organisations. Several countries have introduced policies to ban or limit the use of plastic bags, which could serve as a way of encouraging the public to patronise reusable bags, thereby promoting the reuse principle. While this option has diverse applications and can result in considerable savings, it does have limitations, especially those involving sanitation and public health.

Recycle, which involves reprocessing a material or product, relies on waste that can no longer be reused, but can be reprocessed or transformed into a new material or product. It is enabled by promoting separation at source through policies, incentives and awareness. While it is equally a manufacturing process with environmental impact, this impact is less than that of incineration and landfilling, or that of the production of new products using virgin materials.

There are several benefits to the waste management practice of the 3R principle. These include saving natural resources and reducing the quantity of waste going to the incinerator or the landfill. Opportunities for jobs are equally created in the area of waste collection, sorting, recycling processing and reuse centres. Applying the principle produces less greenhouse gas emissions and air pollution, and also saves energy.

3.2.2.2 *The 6R hierarchy*

This waste management framework is more related to the concept of sustainable manufacturing and plastic-based products (Jawahir and Bradley, 2016; Jayal et al., 2010), as illustrated in Fig. 3.2. At the product level, the complete life cycle of a product consists of main stages that include pre-manufacturing, manufacturing, use and post-use. To achieve sustainable manufacturing and an almost permanent product lifespan, these stages of the product life cycle need to be taken into consideration and a shift needs to be made from the orthodox 3R principle to the innovative 6R principle.

The 6R hierarchy has the options redesign, reduce, remanufacture, reuse, recycle and recovery in descending order of priority. It is a modification of the 3R with the addition of three more options, namely reduce, redesign and remanufacture. The 6R allows for integration across all the stages of the product life cycle and also the theoretical achievement of a closed-loop, multiple life cycle model.

Reduce in the 6R relates to decreasing the usage of resources in pre-manufacturing, materials, energy and others in the process of manufacturing, as well as decreasing waste and emissions in the use stage. Reuse relates to reusing a component or whole product after its initial life cycle for successive life cycles, which enables a reduction in the use of virgin materials to manufacture newer components or products. Redesign relates to reforming new products using recovered materials, components or resources from the former life cycle or generation of the products, and remanufacture refers to reprocessing the product to its original state, or as good as new, and with a safe warranty period without losing functionality.

3.3 Circular economy principles and the 6R framework

The R-framework has been described as the fundamental principle and guideline of the circular economy. Therefore the correlation of the 6R framework waste hierarchy options to the principles of the circular economy by the Ellen MacArthur Foundation was explored. The three fundamental principles driving the circular economy, by design, are to eliminate waste and pollution, circulate products and materials (at their highest value) and regenerate nature (Ellen Macarthur Foundation, 2022). The first principle of eliminating waste and pollution discourages taking raw materials, making them into a product and discarding them as waste. The principle of circulating products and materials (at their highest value) promotes materials being kept in use either as a product or alternatively as components or raw materials. The principle of regenerative nature backs natural processes and enables nature to flourish. Similarly, the 6R framework options are redesign, reduce, remanufacture, reuse, recycle and recovery.

A correlation analysis of the six waste management options of the 6R framework and the principles of the circular economy by the Ellen Macarthur Foundation shows that the waste management options align with or can be categorised under each principle of the circular economy, with the exception of the principle of regenerate nature, as represented in Fig. 3.3. This principle relates to the biosphere, but the focus here is more on the technosphere.

The waste management options of redesign and reduce align with or are enabled by the first principle of eliminating waste and pollution. The options of remanufacture and reuse can be associated with only the second principle of circulate products and materials (at their highest value). Likewise, the options of recycle and recovery align only with the first principle of eliminating waste and pollution. Overall, the analysis shows that the 6R framework and the circular economy principles share a common philosophy, focusing on managing waste by sustainably designing, maximally using and reusing, and recycling so as to improve resource efficiency, reduce plastic waste and advance environmental sustainability.

Furthermore, the integration and application of the 6R framework to the life cycle of plastics will effectively and sustainably achieve a broad and comprehensive circular plastic economy. The 6R hierarchy, with the highest number of most recognised waste management options in the waste hierarchy for achieving circularity, allows for the integration of these options across the life cycle of a product. Table 3.1 exemplifies the various solutions and actions

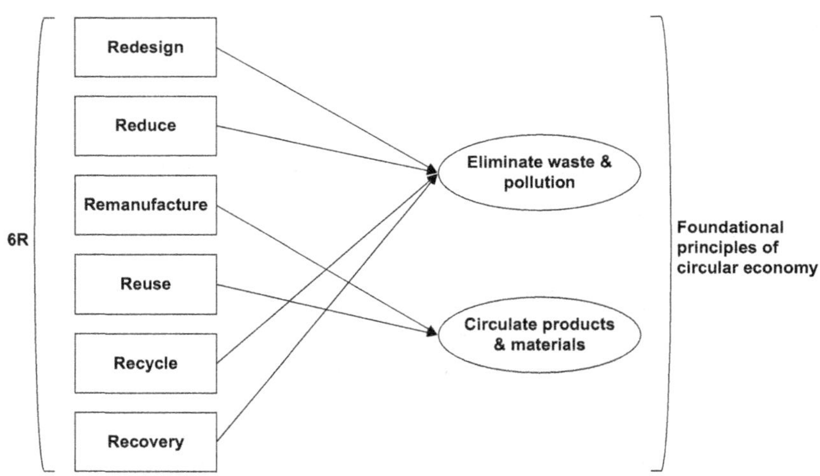

FIGURE 3.3 Correlation between circular economy principles and 6R framework. 6R framework enabling circular economy principles.

TABLE 3.1 Plastic circularity solutions and actions associated with the 6R options.

Plastics life cycle	6R options	Circularity solutions and actions
Production	Redesign	• Innovate and design plastic products for longevity
	Reduce	• Less material in production, e.g. light weighting by design
	Remanufacture	• Providing new life to products, making them as good as new and with the same warranty period • Use of dismantled spare parts for the repair of in-use products
	Reuse	• Re-use of production waste • Re-use of product parts
Use	Reduce	• Buying only the needful • Refillable/reusable containers and bags
	Reuse	• Trade in second-hand plastic goods • Donation to communities • Repurposing plastic products; turning plastic waste into artworks
End-of-life	Recycle	• Quality plastic recyclate • More use of recyclate in plastic product manufacturing • Purchase of plastic products made from recyclate
	Recover	• Shredding • Incineration • Gasification • Pyrolysis • Energy

with respect to the applications of the 6R framework options for advancing circularity in the life cycle of plastics.

Promoting circularity through redesign can take place at the production phase of the life cycle of the product. Redesigning should enable products to be lightweight, durable, reusable, repairable, recyclable and easily disassembled. This can be realised by manufacturers redesigning plastic products for longevity. The design of a product within its production phase influences its longevity (Jiao et al., 2024), use and suitability for a circular economy. This translates into designing and manufacturing plastic materials and products that can withstand wear and tear for a longer period and can equally be reintroduced into the plastic production cycle after use. Redesign is also an important concept in remanufacturing, as it directly affects products that are remanufactured (Du et al., 2013). It contributes to remanufacturing by making the process easier or enabling the capability for remanufacture to take place.

Reduce can take place in both the production and use phases of the life cycle of plastics, where less material can be used in the production of plastic products and where consumers only buy what they need, or reuse and repurpose plastic products. Reduction in plastic use has been achieved in cases where levies are placed on plastic bags in some jurisdictions. Studies have shown how the introduction of a fee on plastic bag consumption in some countries has led to a reduction in plastic bag sales (Dikgang et al., 2012). Achieving circularity through reduce can be realised by reducing parts of products or items made with plastics; buying only what is needed; investing in refillable or reusable food containers, bottles and cups to reduce single-use plastics in terms of takeaway foods and coffee cups; refilling by bulk dispenser, such as bulk containers for foods or kegs; and shopping for groceries with personal reusable bags. Overall, reducing plastics or plastic waste will greatly reduce the adverse impact on the environment and advance circularity.

Remanufacture can be achieved in the production phase of the life cycle of plastics. It can be realised for the advancement of plastic circularity in cases where products are restored to their original state or transformed into a new form by giving new life to many parts of a product with minimal material input. It is also carried out with disassembled parts, where conserved or dismantled spare parts are used for the maintenance or repair of products that are still in use. There are equally instances where products are reconditioned or refurbished instead of remanufactured as a result of the intensive processes and economic viability of remanufacturing (King et al., 2006).

Reuse can be associated with the production and use phase of the life cycle of plastics, where reuse for original purpose, repurpose, repair, sales or donation of used products is promoted. In the production phase, providing opportunities for the collection and reuse of production waste is a key priority for addressing circularity at the production level. Also, realising circularity in the system through reuse is exemplified through trades in second-hand plastic goods such as toys, furniture, plates and several other products in most developing economies; donations to communities; reusing or repurposing plastic bottles or yoghurt tubs as plant pots; reusing used toothbrushes as scrubs; reusing old plastic food or water containers for other storage purposes; turning used plastics or plastic waste into an art piece; and reusable packaging options such as returnable bottles in countries with reuse infrastructure and shared use.

Recycle is related to the end-of-life phase of the life cycle of plastics, which helps limit the volume of plastic waste going to landfills. Recycling solutions for advancing circularity in the plastic system include improved processes in waste collection, separate waste collection and sorting; producing quality plastic recyclate or recycled plastics; manufacturing plastic products from plastic recyclate; purchasing plastic products made from recycled plastic by government entities as a motivation; choosing plastic products that are recyclable where possible and upcycling plastic bags into reusable bags for groceries.

Recovery can equally be associated with the end-of-life phase of the life cycle of plastics, where plastic waste that cannot be recycled is set aside for energy recovery, or disassembled and then used in successive life cycles of the plastic product. Solutions relating to recovery include shredding, incineration, gasification, pyrolysis and energy from plastic waste.

3.4 Consequences of mismanaged plastics

Mismanaged plastics are plastics that are managed ineffectively within the plastics value chain, with their waste landfilled or littered rather than recycled or recovered for energy. They include plastic waste not properly collected, illegally dumped in open spaces, disposed of in unsanitary or uncontrolled landfills, or burnt openly. They arise when circularity is not advanced throughout the life cycle of plastics or across the plastics value chain, where plastics are not sustainably designed, reduced, remanufactured, maximally used, reused, recycled or recovered. These plastics constitute a major environmental challenge with potential severe consequences for the environment and biodiversity. Mismanaged plastics in the economy are known to leak into the environment and subsequently break down into microplastics.

3.4.1 Plastic leakage

Mismanaged plastics result in plastic leakage, thereby causing plastic pollution. Described as the amount of plastics entering the terrestrial environment and water bodies, plastic leakage includes littering, open dumping, unmanaged landfills and accidental release. The leakage of plastics into the environment has adverse impacts on the ecosystem and human health. It comes from several sources and can occur at all stages of the plastics value chain, from the production of plastics to waste generation and disposal.

Plastics have been documented to be on beaches, in basins, lakes and rivers, as well as the soil and human bodies (Organisation for Economic Co-operation and Development, 2022). This leakage exposes human beings and animals to chemicals and additives used in producing plastics, leading to possible health challenges for the reproductive organs, genes and other parts of the human system. The magnitude of plastic leakage into the aquatic environment depends on several factors, including characteristics of the plastic such as weight, proximity of the water body, heat, and weather.

The estimates of the amounts of plastics leaking into the environment have been based largely on assumptions. Globally, about 19–23 million tonnes of plastic waste are estimated to leak annually into the aquatic environment, causing plastic pollution in rivers, lakes, seas and oceans (United Nations Environment Programme, 2024).

3.4.2 Microplastics

Mismanaged plastics are a growing source and concern of microplastic pollution. Microplastics are small plastics in the range of 1 μm to 5 mm in size. They are either produced for use or arise through the weathering or breakdown of plastic products that leak into the ocean (which is a major source) (Andrady, 2011).

Microplastics contain chemicals and additives that are potentially dangerous. They can also be physically dangerous in terms of obstructing or causing physical damage to the feeding organs of organisms when ingested or may attach to the surface or body of the organisms. Ingestion can occur through the cell membranes, gills in ventilation, or direct or indirect exposure paths (Joint Group of Experts on the Scientific, 2015). Due to their small size, they can be ingested by either mistaking them for food, direct attraction or indirectly feeding on food containing microplastics. Microplastics ingested in the tissues and organs of organisms can result in suffocation, intestinal blockage and particle toxicity, together with fibrosis and inflammation (Joint Group of Experts on the Scientific, 2015).

Microbeads and textiles are two major sources of microplastics. Microbeads are microplastics related to personal care products and cosmetics. They can be lost in water and wastewater. However, they contribute minimally to plastic pollution. Nylon, polyester and acrylic are textile plastic fibres that break into microfibres during washing. They have become a foremost source of microplastics in aquatic environments over time (Napper and Thompson, 2016). Additionally, microplastics enter the natural environment through vehicle tyre abrasion. This is influenced by factors such as weather, driving style, road characteristics and composition of tyres.

3.5 Chapter summary

The management of plastic waste for sustainability was discussed in this chapter. Options higher up the waste management hierarchy should be promoted in accordance with the principle of the circular economy. Specifically, the chapter covered the management of plastics and plastic waste in the environment, the waste management hierarchy systems, the relationship between the waste hierarchy principles and the circular economy principles, and the consequences of mismanaged plastics, which include leakages of plastics and microplastics.

The production, use and end-of-life of plastics can be linked to several problems, such as greenhouse gas emissions during production, unsustainable resource use and environmental pollution. The increased production and use of plastic, combined with the poor management of the corresponding waste, have resulted in accumulated plastic in landfills, in oceans and on land and have had negative impacts on human health, biodiversity and the environment.

The waste management hierarchy systems promote the circular economy and a sustainable environment. This ensures sustainable production, consumption and waste management, leading to a reduced impact on the environment. Therefore, it is imperative and valuable to ensure that the use and application of plastics align globally with plastic circularity and are realised with the application of the waste hierarchy options.

The Waste Framework Directive 2008/98/EC, introduced in 2008, and the amendments to the directive in 2018, are an essential component of waste legislation in the EU. It articulates principles for reducing environmental damage from improper and unsustainable waste management. Described as the milestone of modern waste management, this directive clearly defines the priority order of the five options or measures of the waste hierarchy, namely prevention, preparing for reuse, recycling, other recovery (e.g. energy recovery) and disposal. It advocates for a priority order from the most desirable option of prevention at the top of the hierarchy to the least desirable option of disposal at the bottom. The hierarchy supports the transition of the EU to a circular economy, with a progressive decrease in the mass of waste going to landfill. By following the top-down priority guidelines, countries are expected to design and implement plans and policies to promote and enable a shift towards the options higher up the waste hierarchy.

Another category of waste management hierarchical options, termed the R framework system, was equally analysed. It relates to the EU waste hierarchy system, both in theory and in practice. The Rs in the waste hierarchy are the options, in descending order of preference or priority, of refuse, rethink, reduce, reuse, repair, refurbish, remanufacture, repurpose, recycle and recover. The R framework is apparently the most comprehensive, providing the highest number of options in the waste hierarchy for the achievement of the circular economy in the plastics value chain.

The circular economy shifts away from the linear model of extraction, production, consumption and eventual deposit of waste to a more efficient and sustainable use of resources and elimination of waste, based on the application of the various options higher up the waste management hierarchy. A correlation analysis of the waste management options of the 6R framework and the principles of the circular economy by the Ellen Macarthur Foundation shows that the waste management options align with all the principles of the circular economy. The analysis also shows that 6R framework and the circular economy principles share a common philosophy, focusing on managing waste by sustainably designing, maximally using and reusing, and recycling so as to improve resource efficiency, reduce plastic waste and advance environmental sustainability. In advancing the circularity of plastics across their life cycle, various solutions and applications of the 6R framework options were exemplified.

Furthermore, the consequences of mismanaged plastics were examined, as they are known to result in plastic leakages and microplastics. This has adverse impacts on the ecosystem and human health. Microplastics are a major form of plastic loss to the environment. They contain chemicals and additives that are potentially dangerous. Therefore broader plastic frameworks are required to change direction regarding the general use and application of plastics and the management of plastic waste. This should promote options higher up the waste management hierarchy.

References

Ahmadi, M., 2017. Evaluating the Performance of 3Rs Waste Practices: Case Study-Region One Municipality of Tehran. Advances in Recycling & Waste Management 02 (02). Available at https://doi.org/10.4172/2475-7675.1000130.

Andrady, A.L., 2011. Microplastics in the marine environment. Marine Pollution Bulletin 62 (8), 1596–1605. Available at https://doi.org/10.1016/j.marpolbul.2011.05.030.

Berger, R., 2023. The plastic waste management framework. White paper for Alliance to End Plastic Waste.

Committee of the Regions, 2015. European Commission COM/2015/0614: Communication from the Commission to the European Parliament, the Council, the European Economic and Social Committee and the https://eur-lex.europa.eu/resource.html?uri=cellar:8a8ef5e8-99a0-11e5-b3b7-01aa75ed71a1.0012.02/DOC_1&format=PDF.

Council Directive of 15 July 1975 on waste (75/442/EEC) 1975 http://data.europa.eu/eli/dir/1975/442/2006-05-17 https://eur-lex.europa.eu/legal-content/en/TXT/?uri=CELEX:31975L0442.

Dikgang, J., Leiman, A., Visser, M., 2012. Analysis of the plastic-bag levy in South Africa. Resources, Conservation and Recycling 66, 59–65. Available at https://doi.org/10.1016/j.resconrec.2012.06.009.

Du, Y., Cao, H., Chen, X., Wang, B., 2013. Reuse-oriented redesign method of used products based on axiomatic design theory and QFD. Journal of Cleaner Production 39, 79–86. Available at https://doi.org/10.1016/j.jclepro.2012.08.032.

Ellen Macarthur Foundation, 2022. What is a circular economy? Ellen MacArthur Foundation.

Ghisellini, P., Cialani, C., Ulgiati, S., 2016. A review on circular economy: The expected transition to a balanced interplay of environmental and economic systems. Journal of Cleaner Production 114, 11–32. Available at https://doi.org/10.1016/j.jclepro.2015.09.007.

Giusti, 2009. A review of waste management practices and their impact on human health. Waste Management 29 (8), 2227–2239. Available at https://doi.org/10.1016/j.wasman.2009.03.028.

Jawahir, I.S., Bradley, R., 2016. Technological elements of circular economy and the principles of 6r-based closed-loop material flow in sustainable manufacturing. Procedia CIRP 40, 103–108. Available at https://doi.org/10.1016/j.procir.2016.01.067.

Jayal, A.D., Badurdeen, F., Dillon, O.W., Jawahir, I.S., 2010. Sustainable manufacturing: Modeling and optimization challenges at the product, process and system levels. CIRP Journal of Manufacturing Science and Technology 2 (3), 144–152. Available at https://doi.org/10.1016/j.cirpj.2010.03.006.

Jiao, H., Ali, S.S., Alsharbaty, M.H.M., Elsamahy, T., Abdelkarim, E., Schagerl, M., et al., 2024. A critical review on plastic waste life cycle assessment and management: challenges, research gaps, and future perspectives. Ecotoxicology and Environmental Safety 271. Available at https://doi.org/10.1016/j.ecoenv.2024.115942.

Joint Group of Experts on the Scientific Aspects of Marine Environmental Protection) Sources, fate and effects of microplastics in the marine environment: a global assessment. (2015).

King, A.M., Burgess, S.C., Ijomah, W., McMahon, C.A., 2006. Reducing waste: Repair, recondition, remanufacture or recycle? Sustainable Development 14 (4), 257–267. Available at https://doi.org/10.1002/sd.271.

Napper, I.E., Thompson, R.C., 2016. Release of synthetic microplastic plastic fibres from domestic washing machines: Effects of fabric type and washing conditions. Marine Pollution Bulletin 112 (1-2), 39–45. Available at https://doi.org/10.1016/j.marpolbul.2016.09.025.

Organisation for Economic Co-operation and Development, 2022. Global plastic waste set to almost triple by 2060, says OECD. OECD, 12.

Pires, A., Martinho, G., 2019. Waste hierarchy index for circular economy in waste management. Waste Management 95, 298–305. Available at https://doi.org/10.1016/j.wasman.2019.06.014.

Potting, J., Hekkert, M.P., Worrel, E., Hanemaaijer, A., 2017. Circular Economy: Measuring Innovation in The Product Chain. PBL Netherlands Assessment Agency.

United Nations Environment Programme, 2024. Plastic pollution. United Nations Environment Programme.

Van Buren, N., Demmers, M., Van der Heijden, R., Witlox, F., 2016. Towards a Circular Economy: The Role of Dutch Logistics Industries and Governments. Sustainability 8 (7), 647. Available at https://doi.org/10.3390/su8070647.

Yan, J., Feng, C., 2014. Sustainable design-oriented product modularity combined with 6R concept: A case study of rotor laboratory bench. Clean Technologies and Environmental Policy 16 (1), 95–109. Available at https://doi.org/10.1007/s10098-013-0597-3.

Zhang, C., Hu, M., Di Maio, F., Sprecher, B., Yang, X., Tukker, A., 2022. An overview of the waste hierarchy framework for analyzing the circularity in construction and demolition waste management in Europe. Science of the Total Environment 803. Available at https://doi.org/10.1016/j.scitotenv.2021.149892.

Chapter 4

Impact and sustainability of plastic

4.1 The impact of plastics and the life cycle assessment

Plastic is a very important material and has great properties that are useful for many applications. However, the continued increase in the global production and use of the material comes with corresponding consequences and impact on the environment, such as resource depletion, waste accumulation and ecological problems. These challenges associated with the production, use and end-of-life phases of the life cycle of plastic, particularly single- or short-term use plastics, mean that the material is generally regarded as a pollutant to the environment, with devastating impacts. This situation has continued to raise concerns and receive attention within and outside the industry.

The various impacts of plastics on the environment throughout their life cycle need to be examined and evaluated in order to enable and promote efficiency in resource use, conservation of energy, reduction of plastic waste, circularity of plastics and sustainability of the environment. This evaluation is performed through the life cycle assessment (LCA). The LCA of plastics is a scientific approach to quantifying the impact of plastic on the environment over its entire lifetime, including activities such as raw material extraction, polymer production, product manufacturing, transportation, consumption and waste management (Guinée, 2002). It encompasses a wide range of societal issues, taking into consideration the natural environment, resources and human health. It has been found to be useful in policy-making, product development, process development and strategic planning. The assessment is guided by ISO 14040:2006, which describes the principles and framework for LCA.

The LCA of plastics contains key elements: (1) Goal and scope definition, (2) inventory analysis, (3) impact assessment and (4) interpretation (Bisinella et al., 2021). In this case, LCA can be used to assess, systematically, the environmental impacts of the plastic material–based product system from raw material acquisition to final disposal, in accordance with the stated goal and scope. The scope of analysis can be as follows:

- Cradle-to-grave. The life cycle stages from the extraction or acquisition of raw materials to manufacturing, use and end-of-life options.

Plastic Circularity. DOI: https://doi.org/10.1016/B978-0-443-40667-6.00006-8

- Cradle-to-gate. The life cycle stages from the extraction or acquisition of raw materials to the point at which the product leaves the organisation undertaking the assessment.
- Gate-to-gate. From input to the point at which the product leaves the organisation undertaking the assessment.

When the goal of an LCA is defined, this is used to derive the functional unit. A functional unit is the defined performance of a product system or a service provided by a product choice and this is used as a reference unit for comparing impacts.

The second and inventory stage of the analysis involves identifying and quantifying the associated resources in terms of energy, materials and emissions. The inventory is then used to assess the potential impact of the loads on the environment based on emission factors. Evaluation of the results and available options or opportunities helps to identify hot spots and reduce the environmental impacts (Eea, 2023). These analyses are supported by confidence and sensitivity analysis. The successful implementation of plastics LCA depends on the availability and quality of the life cycle inventory (LCI) data and the impact assessment method. A comprehensive plastics LCA also ensures that environmental problems are not shifted from one particular stage of the life cycle to another.

The environmental impacts usually quantified comprise multiple categories, including global warming, carcinogens, non-carcinogens, ozone layer depletion, aquatic ecotoxicity, aquatic eutrophication, terrestrial ecotoxicity, respiratory inorganics, respiratory organics, aquatic acidification, terrestrial acidification/nitrification, non-renewable energy (primary) and land occupation (Horowitz et al., 2018). The life cycle environmental impacts of plastics are regarded as substantial, with an estimated greenhouse gas (GHG) emission generation of 1.781 Gt CO_2 annually throughout the life cycle (Zheng and Suh, 2019). This entails 1.085 Gt of production, 0.535 Gt of conversion and 0.161 Gt of end-of-life. The generation represents more than 3% of total emissions globally.

4.1.1 Plastic manufacturing and the environmental impact

Plastics are produced mainly from fossil fuels such as oil, natural gas and coal, and they consist of numerous elements such as nitrogen, carbon, oxygen, chloride, hydrogen and chloride (Wasewar et al., 2022). These are non-renewable resources with associated impact. The production of plastics involves the processes of extracting and purifying crude oil to derive the raw material. The process of extraction involves large-scale operations such as mining, drilling and hydraulic fracturing. These various operations and activities are energy- and resource-intensive and have substantial impacts on the environment, including plastic pollution, habitat destruction, GHG emissions and marine contamination.

In manufacturing plastic products, various pellets are produced and then further processed. Plastic pellets derived from petroleum are produced through the polymerisation of oil monomers. These pellets are mixed with various additives to produce different plastic products. Several chemical compounds are added in the process of moulding for the plastics' performance, functionality and ageing. These processes also have an impact on the environment due to energy being consumed and contaminants being emitted into the atmosphere. In addition, quantities of pellets are sometimes lost by accident, as a result of leakage into the environment during the production of plastics. This leakage can be difficult to trace. Plastic pieces that are produced during different processes of production also cause contamination.

Furthermore, the transportation of processed and unprocessed materials between extraction sites, raw material markets, refineries and factories for the manufacture of the final product needs to be examined for environmental impacts. All transportation activities and means are exposed to material losses through leaks that may result in accidents and explosions. Risks, including health, safety and environmental impacts, also need to be taken into consideration during transportation (Williams and Rangel-Buitrago, 2022).

4.1.2 Plastic use contributes to environmental impact

The single- or short-term use of plastics often results in the accumulation of plastic waste, which impacts the environment negatively. Many plastic products can be subjected to intensive use and reused within the plastic consumption stage for better environmental outcomes. LCA studies have shown that the reuse of plastic will both reduce the quantity of plastic waste going to the landfill and lessen the impact on the environment. Considering the quantity of plastics being produced yearly, researching and advancing the reuse of plastic is crucial so as to reduce the accumulation of plastic waste and environmental deterioration.

4.1.3 End-of-life impact of plastics

The challenge of waste mismanagement largely involves the inefficient and unsustainable means of plastic waste disposal. The majority of plastic waste ends up in landfills, with relatively little being recycled or incinerated. This plastic waste often becomes contaminated in landfills, making it inappropriate for recovery. The lasting presence of plastic waste in the landfill contributes to the build-up and growth of persistent pollutants. Their inability to degrade has a negative impact on the ecosystem, human health and water bodies. Plastic waste that is buried for a long time has negative ecological impacts, such as decreased microbial diversity, altered physicochemical composition and deteriorated soil fertility (Ali et al., 2023).

Owing to the various challenges of plastic waste, plastic recycling has attracted a great deal of attention lately as plastic waste continues to increase. The LCA of plastic has shown that recycling produces more desirable environmental outcomes than incineration and landfill. Recycling plastic is essential for eliminating pollution, preserving resources and lessening the negative impact of plastic waste on the environment. It contributes to decreasing energy use, extending the life of landfill sites and lowering GHG emissions. It also has the added advantage of creating employment and advancing plastic circularity. However, contaminating plastic recyclables with residual waste may hinder recycling. In addition, the recycling of plastic waste can be an expensive undertaking requiring significant investments in water, electricity, technology, infrastructure and water (Gopinath et al., 2020). Furthermore, each recycling cycle may result in lower-quality plastic recyclate, and in order to restore desired qualities, newly produced plastics or additives are frequently mixed in.

While plastic recycling is extremely important in limiting the impact on the environment and preserving resources, addressing problems related to cost, contamination and quality degradation is essential to optimise the effectiveness and performance of the recycling process. Plastic recyclate or recycled plastic may come with drawbacks, such as lower product quality in comparison to products made from virgin polymers.

4.1.4 Gaps in life cycle assessment of plastics

A number of limitations have been identified in the application of LCA of plastics. Firstly, it is generally argued that there is currently no publicly available tool to help gather reliable LCI for plastics (Beylot et al., 2020). This problem becomes more intense when trying to create an LCI for a unique plastic material or product without publicly available background data. When using pre-existing LCI data, compatibility along technological, temporal and geographical boundaries could pose additional challenges, since most of these studies only provide results for specific parameters. Having the ability to modify or vary these parameters would greatly improve the process of making assessments and decisions.

In addition, while the market for biodegradable plastics is expanding, the majority of plastic LCAs have concentrated on non-biodegradable plastics derived from fossil fuels (Jiao et al., 2024). Even though some studies examined both fossil fuel and bio-plastics, fundamental characteristics of plastic pollution have been overlooked since the focus has mostly been on the improvement of material circularity. Important elements such as plastic shape, size, or density need to be taken into consideration as they contribute fundamentally to the fragmentation, fate, exposure and transport of plastics within the environment and the associated impacts. Progress is, however, apparently being made to incorporate these elements within LCA methodologies.

Several LCAs of plastics exclude the end-of-life stage, creating the impression that plastics are the most preferred material choice despite the fact that plastics may continue to deteriorate and release toxic compounds into the environment. Where LCAs have been done for end-of-life, most of these studies examined the recycling alternative of plastic waste management instead of other alternatives.

Another limitation is that most of the major indicators used in characterising environmental impacts from plastic material extraction, manufacturing, transportation and other related processes and activities with respect to the life cycle of plastics do not capture the damaging impact of plastics when lost to the environment. Relevant and important impacts such as human toxicity, biodiversity loss or ecosystem toxicity are not included. The exclusion of these indicators or impacts from the assessment possibly implies that these environmental impact indicators could have been accounted for and assessed within other indicators.

Furthermore, there is the challenge of accounting for additives, which are a critical component of plastic products. Additives need to be accounted for in order to obtain accurate assessment results, as they alter the characteristics of plastic products. More still needs to be done regarding their inclusion in LCA, as their coverage is still inadequate (Bishop et al., 2021). Integrating these compounds into plastic LCA is challenging, as many of them cannot be captured by a general LCA. Similarly, additives can affect the end-of-life management alternatives of plastics. Considering that the end-of-life of plastic is rarely accounted for in LCA, quantifying the impact of additives on the environment can be problematic.

4.2 Making plastics more sustainable

This current pattern of plastic applications and flow, which involves the production, use and generation of plastic waste, is unsustainable, as it has adverse impacts on the environment. Plastics need to be made more sustainable by advancing sustainability practices and activities throughout the life cycle of plastics. These practices will contribute to the reduction of plastic waste, pollution and GHG emissions. Fig. 4.1 illustrates the sustainability and unsustainability cycles or pathways for the plastics life cycle. It shows different actions towards promoting and enabling more sustainability across the life cycle of plastics. The unsustainability pathway shows the different impacts (inward arrows) of each stage on the environment, and the sustainability pathway indicates the different benefits (outward arrows) of each stage of the plastic system.

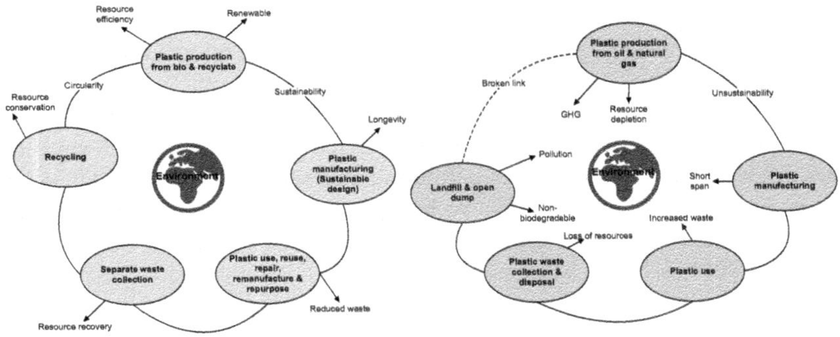

FIGURE 4.1 Sustainability and unsustainability in the life cycle of plastics.

For plastic to be completely sustainable, it must be economically, socially and environmentally sustainable. It must contribute positively to these categories of sustainability. Economic sustainability has to do with the contribution to the overall economy of a place or country. Social sustainability relates to education, awareness, jobs and public participation. Environmental sustainability involves contributing to resource and energy savings and reducing environmental impacts.

4.2.1 Economic sustainability of plastics

Plastic makes significant contributions to the global economy. In 2020 the annual production cost of plastics globally was valued at $113 billion, and the plastic recycling industry was estimated at $33 billion, with a projection of $47.3 billion by 2026 (Wood, 2020). Plastic has a low cost of production, as products made from plastics are cost-effective, considering that single moulded components can be made into various complex shapes. The economic sustainability of plastic can be related to a means of growing the economy without depleting or damaging the natural resources for the production of plastics. It describes a system in which resources for plastic production are consumed in a way that is not detrimental to the environment or the needs of current or future generations. It involves practices and actions that support and promote the long-term economic growth of plastics without impacting the environmental and social aspects of society negatively. Plastic's economic sustainability seeks to balance the economic growth of the material with environmental responsibility and social equity. It deters a strong plastics economy that comes at the expense of public health, social well-being or the environment.

In practice, the plastics economy can become more sustainable by promoting certain actions that include reducing the reliance on fossil fuels for the production of plastics and promoting economic models that are centred on the circular economy. Less reliance on fossil fuels such as oil and natural gas will entail shifting to renewable sources such as bio-based plastics, which contributes to making the economy more sustainable. The promotion of the circular plastic economic model will also involve designing plastic products to be reused and repaired and increasing plastic recycling at the end-of-life of the product. This enables the reduction of plastic waste and exploitation of resources, thereby saving plastic resources and contributing to making the economy of plastic more sustainable. Some of the factors that influence the economic sustainability of plastics include the responsible management of resources, innovation and efficiency capacity of the economic system and policies and programmes.

4.2.2 Social sustainability of plastics

The plastics industry can be socially responsible by offering social inclusivity and supporting the well-being of the public and society. The social sustainability of plastics should ensure that the growth of plastics promotes equity, participation of the public, access to education and career opportunities and provision of healthcare. All these help to build inclusive societies, reduce inequality, preserve social cohesion and enable access to plastic resources. The demonstration of plastics in advancing social sustainability in health care could be seen during the COVID-19 pandemic, where plastic PPE played a crucial role in protecting and saving lives, as well as contributing towards ending the pandemic. Plastic products are used in surgery, pharmaceuticals, medical packaging and drug delivery, among other things. Plastics also contribute to safety, where their application in cars helps to reduce weight; plastic components in helmets protect the lives of motorbike riders and cyclists. Social sustainability of plastic can be achieved through seeking the growth of plastics while simultaneously promoting policies and practices that push for the participation and inclusivity of the public, access to education, improvement of health and well-being of people and the creation of safe and clean environments.

4.2.3 Environmental sustainability of plastics

Due to several desirable properties of plastics, numerous sectors use plastics, such as packaging, vehicles and construction. Plastics help to make products lighter. The use of plastics in packaging ensures resource efficiency, reduces waste and saves energy consumption, which would have impacted the environment negatively. The subsequent waste from plastic packaging at the end-of-life can be recycled. The use of plastic as a substitute for iron or steel in PVC windows and water pipes in construction, as well as components in vehicles, contributes to the consumption of relatively less energy for production, with relatively less consequent impact on the environment.

Despite these advantages and benefits, the material does have a negative impact on the environment. A shift towards the more sustainable production and consumption of plastics and management of the corresponding waste is required to mitigate this environmental impact. Making the production of plastics more environmentally sustainable involves consuming less material in the manufacture of plastic products, making use of sustainable product design, reducing single-use packaging and products and producing more recycled and bio-based plastic products. Longer and more intensive use, reuse and repair can be practised in consumption. All this helps limit plastic waste generation. The end-of-life management of plastic can be made more sustainable by improving collection, promoting separate waste collection and increasing recycling of plastic waste.

4.3 Plastics within planetary boundaries

All life on earth, including the geosphere, atmosphere, hydrosphere, cryosphere and biosphere, relies on the health of the Earth system. Planetary boundaries are the framework used to define the limits of how much human activity can impact the Earth's system (Rockström et al., 2009). The environment might no longer be capable of self-regulating after these limits are reached. The framework is grounded in empirical evidence showing that human activity has emerged as the primary cause of environmental change globally, particularly in industrialised nations since the period of the Industrial Revolution. Planetary boundaries are characteristics of an Earth system that is interrelated. Deep biophysical alterations have the potential to cause irreparable shifts in the Earth's system, particularly in relation to the two main boundaries of biodiversity loss and climate change (McKay et al., 2022). Within the framework, there is a possibility of contravening planetary boundary thresholds, which could be hazardous or even disastrous.

With the dependence of the global economy on fossil fuels, the quick growth or increase in the production of plastics has worsened the triple planetary problems of waste pollution, habitat loss and GHG emissions. The current volume of plastic production disturbs the Earth'ssystem dynamics mainly because the extraction and use of fossil fuels are connected to the climate system. The direct strain on the climate planetary boundary, which has already been surpassed, increases with increasing production rates. Plastics could become independent of carbon emissions from fossil fuels if there is a shift to bio-based feedstocks (Gerassimidou et al., 2021). This will affect the integrity of the biosphere, biogeochemical fluxes, and land and water systems, which are already beyond the planetary boundaries. The plastic circularity principle has been proposed to address the triple planetary problem of habitat loss, environmental pollution arising from plastic waste and GHG emissions from the production of plastics and end-of-life treatment of the waste (Bachmann et al., 2023). Different circularity measures are emerging for the reduction of plastics GHG emissions, including bio-based plastic production, plastic recycling and carbon capture and utilisation. Table 4.1 presents circularity strategies to address the various planetary problems with regard to the plastic system.

Plastic pollution is a threat to the environment, human health and food security. Globally, the unregulated production of plastics and the rise in plastic waste greatly surpass the mitigation efforts to address the problem of plastic pollution. Increasing emphasis has continued to be placed on recycling to reduce plastic waste and meet the demand for plastics. Likewise, a considerable reduction in the quantity of virgin plastics produced is required for plastic recycling to have a meaningful effect on the existing situation. This reduction must be paired with better policies and sustainable design standards that facilitate the reuse and recovery of plastic material.

Furthermore, the chemical diversity of plastics is another important consideration, as chemical additives used in the modification and enhancement of plastic properties in plastic production are potentially harmful to the

TABLE 4.1 Planetary challenges and circular economy strategies.

Planetary problems	Source	Circularity economy strategies
Environmental pollution	• Plastic waste mismanagement	• Reduced material and production waste • Reuse • Repair • Repurpose • Re-manufacturing • Recycling
Habitat loss	• Extraction and processing of fossil fuels	• Resource efficiency • Bio-based plastics
GHG emissions	• Plastic production • End-of-life treatment	• Production of bio-based plastics • Application of plastic recyclates • Carbon capture and utilisation • Recycling

GHC, Greenhouse gas.

environment. The chemicals' hazardous properties include endocrine disruption, carcinogenicity, developmental toxicity, nerve damage, biocide effects and metabolic effects (Wiesinger et al., 2021). A significant number of plastics contain poorly documented chemical additives and this situation poses a risk of disturbing Earth system processes. Plastics and their chemical additives, such as plasticisers, pigments and flame retardants, are bio-accumulate, persistent and adversely impactful on the environment with respect to the ecosystem level (Huang et al., 2021).

4.4 Sustainability in a circular economy

The unique properties that make plastic desirable for mass production are the same qualities that make it problematic; a case in point here is single-use packaging. Its problematic nature makes plastic more susceptible to improper disposal and mismanagement. For many years, the detrimental impacts of plastic pollution have been publicly acknowledged. However, more recently, people are becoming more conscious of and knowledgeable about the detrimental impacts that plastic pollution has on global issues, such as air pollution, climate change and human health. Taking actions to lessen these impacts while maintaining the usefulness of plastics will become more crucial as the plastic industry develops. Making the transition to a more sustainable and circular plastic economy will be essential to achieving this.

Circularity of plastics, which emphasises the effective use and recirculating and recycling of plastic resources, can support sustainability. However, the sustainability of plastics spans a wide range of objectives and factors, including environmental, economic and social. Realising the sustainability of plastic requires integrating the circular plastic economy principles. Equally, it requires tackling a wider range of challenges associated with plastics in order to guarantee resilience and long-term well-being across diverse aspects of the environment and society.

Furthermore, in the context of sustainability and the circular economy, it is critical to assess the life cycle of plastic products, as the extension of the product life cycle is one of the practical measures targeted at achieving a circular economy (Circular economy in Europe: Developing the knowledge base, 2016). The circular economy has been acknowledged to advance the minimal use of material, resources and energy, as well as the reduction of waste.

As governments and countries continue to set sustainability and circularity goals and targets, plastic product manufacturers and recyclers need to be engaged to develop new sustainable designs and technologies, explore other alternatives to plastics, create solutions that improve the reduction and recycling of plastic waste and foster stakeholder partnerships. Enabling sustainability within a circular economy will involve reducing plastic material inputs, losses, residuals and emissions, while increasing the share of renewables, energy efficiency, value durability of plastic products and the recycling of plastic waste. These practices lessen the burden on the environment. Sustainability problem-solving of plastics should be promoted while considering the economic, social and environmental impact.

4.5 Chapter summary

This chapter dealt with the impacts of plastics and the contribution of plastic to environmental, economic and social sustainability. In particular, the LCA of plastics was discussed, with an examination of the impacts of the different phases of plastic manufacturing, consumption and end-of-life. Enabling plastics to become more sustainable, operating within planetary boundaries and embedding plastic sustainability and circularity were also explored.

Plastic has an impact on the environment throughout its life cycle. From the continuous rise in plastic production, consumption and the eventual mismanagement of the end-of-life, the material has an increased impact on the environment. This situation continues to raise concerns about the life cycle of plastics. Plastic LCA is a technique used to quantify plastic's impact on the environment over its entire lifetime, including activities such as raw material extraction, polymer production, product manufacturing, transportation, consumption and waste management. This assessment promotes efficiency in resource use and conservation of energy, among other benefits. The production of plastics involves the processes of extracting and purifying crude oil to derive the raw material, as well as the transportation of processed and unprocessed materials, all of which are energy- and resource-intensive, thereby making substantial impacts on the environment. Also, the single- or short-term use of plastics results in most of the corresponding waste accumulating in landfills or open dumps, with consequential negative impacts on the environment. Only a small percentage of this waste is being recycled. The inability of plastic waste to degrade harms the ecosystem, human health and water bodies. However, plastics LCA has shown that recycling has more desirable environmental outcomes than incineration and landfill. Regrettably, the application of this LCA of plastics has limitations: reliable life cycle inventory is limited, most assessments are on fossil fuel–produced plastics, and the damaging impact of plastics when lost to the environment is not captured.

The flow of plastics, including its production, use and end-of-life, within the value chain system is currently unsustainable. Making the life cycle of plastics more sustainable can contribute to minimal plastic waste, pollution and greenhouse gas emissions. In becoming more sustainable, plastics must exhibit economic, social and environmental sustainability. Economic sustainability has to do with contributing to the overall economy of a place or country. It entails actions that promote the long-term economic growth of plastics without impacting the environmental and social aspects of society negatively. Social sustainability ensures that the growth of plastics promotes equity, public participation, education and health care, which helps build inclusive societies, reduce inequality and preserve social cohesion in accessing plastic resources. Environmental sustainability involves contributing to resource and energy savings and reducing environmental impacts. Overall, the three pillars of plastic sustainability advocate for the use of less material in the manufacture of plastic products, making use of sustainable product design, producing more recycled and bio-based plastic products, reducing single-use packaging and products, practising longer and more intensive use, reuse and repair, and increasing recycling.

Furthermore, the principle of plastic circularity has been proposed to address the triple planetary problem of habitat loss, environmental pollution arising from plastic waste and GHG emissions from the production of plastics and the end-of-life treatment. Planetary boundaries describe the limits of human activity's impact on the Earth system, where the environment might be incapable of self-regulating after these limits are reached. The global reliance on the continued production of plastics from fossil fuels has worsened the triple planetary problem.

Transitioning to a more sustainable and circular plastic economy requires taking actions to lessen the several impacts on the environment and reduce the triple planetary challenges, while preserving the usefulness of plastics, increasing resource efficiency and reducing waste. Circularity of plastic emphasises the effective use of plastics and recycling of plastic waste, and can support sustainability. Sustainability of plastics spans a wide range of objectives and factors, including environmental, economic and social. Achieving plastic sustainability requires integrating the circular plastic economy principles. This involves the continuous availability of the plastic resource while tackling a wider range of challenges associated with plastic's impact on the environment. Likewise, enabling sustainability within a circular economy will involve reducing plastic material inputs, losses, residuals and emissions, while increasing the share of renewables, energy efficiency, value durability of plastic products and the recycling of plastic waste.

References

Ali, S.S., Elsamahy, T., Zhu, D., Sun, J., 2023. Biodegradability of polyethylene by efficient bacteria from the guts of plastic-eating waxworms and investigation of its degradation mechanism. Journal of Hazardous Materials 443 (Pt B), 130287. Available at https://doi.org/10.1016/j.jhazmat.2022.130287.

Bachmann, M., Zibunas, C., Hartmann, J., Tulus, V., Suh, S., Guillén-Gosálbez, G., et al., 2023. Towards circular plastics within planetary boundaries. Nature Sustainability 6 (5), 599–610. Available at https://doi.org/10.1038/s41893-022-01054-9.

Beylot, A., Corrado, S., Sala, S., 2020. Environmental impacts of European trade: interpreting results of process-based LCA and environmentally extended input–output analysis towards hotspot identification. The International Journal of Life Cycle Assessment 25 (12), 2432–2450. Available at https://doi.org/10.1007/s11367-019-01649-z.

Bishop, G., Styles, D., Lens, P.N.L., 2021. Environmental performance comparison of bioplastics and petrochemical plastics: A review of life cycle assessment (LCA) methodological decisions. Resources, Conservation and Recycling 168, 105451. Available at https://doi.org/10.1016/j.resconrec.2021.105451.

Bisinella, V., Christensen, T.H., Astrup, T.F, 2021. Future scenarios and life cycle assessment: systematic review and recommendations. International Journal of Life Cycle Assessment 26 (11), 2143–2170. Available at https://doi.org/10.1007/s11367-021-01954-6.

EEA, 2023. Life cycle assessment. European Environment Agency 20.

European Environment Agency, 2016. Circular Economy in Europe: Developing the Knowledge Base (EEA Report No. 2/2016). Publications Office of the European Union.

Gerassimidou, S., Martin, O.V., Chapman, S.P., Hahladakis, J.N., Iacovidou, E., 2021. Development of an integrated sustainability matrix to depict challenges and trade-offs of introducing bio-based plastics in the food packaging value chain. Journal of Cleaner Production 286. Available at https://doi.org/10.1016/j.jclepro.2020.125378.

Gopinath, K.P., Nagarajan, V.M., Krishnan, A., Malolan, R., 2020. A critical review on the influence of energy, environmental and economic factors on various processes used to handle and recycle plastic wastes: Development of a comprehensive index. Journal of Cleaner Production 274. Available at https://doi.org/10.1016/j.jclepro.2020.123031.

Guinée, J.B., 2002. Handbook on life-cycle assessment operational guide to the ISO Standards. The International Journal of Life Cycle Assessment 7, 311–313. Available at https://doi.org/10.1007/BF02978897.

Horowitz, N., Frago, J., Mu, D., 2018. Life cycle assessment of bottled water: A case study of Green2O products. Waste Management 76, 734–743. Available at https://doi.org/10.1016/j.wasman.2018.02.043.

Huang, Y., Li, W., Gao, J., Wang, F., Yang, W., Han, L., et al., 2021. Effect of microplastics on ecosystem functioning: Microbial nitrogen removal mediated by benthic invertebrates. Science of The Total Environment 754, 142133. Available at https://doi.org/10.1016/j.scitotenv.2020.142133.

Jiao, H., Ali, S.S., Alsharbaty, M.H.M., Elsamahy, T., Abdelkarim, E., Schagerl, M., et al., 2024. A critical review on plastic waste life cycle assessment and management: Challenges, research gaps, and future perspectives. Ecotoxicology and Environmental Safety 271. Available at https://doi.org/10.1016/j.ecoenv.2024.115942.

McKay, D.I.A., Staal, A., Abrams, J.F., Winkelmann, R., Sakschewski, B., Loriani, S., et al., 2022. Exceeding 1.5°C global warming could trigger multiple climate tipping points. Science (New York, N.Y.) 377 (6611). Available at https://doi.org/10.1126/science.abn7950.

Rockström, J., Steffen, W., Noone, K., Persson, A., Chapin, F.S., Lambin, E., et al., 2009. Planetary boundaries: Exploring the safe operating space for humanity. Ecology and Society 14 (2). Available at https://doi.org/10.5751/ES-03180-140232.

Wasewar, K.L., Kumar, S., Pal, D., Uslu, H., 2022. Life Cycle Assessment (LCA) of Plastics. Plastic and Microplastic in the Environment: Management and Health Risks. Wiley, India, pp. 245–263. Available at https://doi.org/10.1002/9781119800897.ch13.

Wiesinger, H., Wang, Z., Hellweg, S., 2021. Deep Dive into Plastic Monomers, Additives, and Processing Aids. Environmental Science & Technology 55 (13), 9339–9351. Available at https://doi.org/10.1021/acs.est.1c00976, https://doi.org/10.1021/acs.est.1c00976 Available from.

Williams, A.T., Rangel-Buitrago, N., 2022. The past, present, and future of plastic pollution. Marine Pollution Bulletin 176, 113429. Available at https://doi.org/10.1016/j.marpolbul.2022.113429.

Wood, L., 2020. Global plastic recycling market report 2021: Market to reach $47.3 billion by 2026 from $33 billion in 2020. Research and Markets.

Zheng, J., Suh, S., 2019. Strategies to reduce the global carbon footprint of plastics. Nature Climate Change 9 (5), 374–378. Available at https://doi.org/10.1038/s41558-019-0459-z.

Chapter 5

Data building in the plastics value chain

5.1 Role of data in the value chain

Data associated with the plastics value chain – including all the value chain stages of plastic material production, plastic product manufacturing, plastic product use and waste generation, plastic waste collection and handling, plastic waste sorting, plastic waste recycling and plastic recyclate market – should be captured. This is because the processing and utilisation of data are important in facilitating the application of circularity strategies towards realising circularity and sustainability across the plastics value chain. Efficiently managing and monitoring this data in the entire life cycle of plastic is crucial to achieving this objective. To advance circularity strategies such as reduce, reuse, repurpose, remanufacture and recycle in the value chain system, data is required to be complete, reliable, accessible and integrated (Berglund-Brown et al., 2022).

The building of data within the plastics value chain represents an essential tool in the hands of stakeholders to integrate circular and sustainability thinking into the decision- and policy-making process, particularly at the national level. This information is useful to the government and non-governmental organisations in assessing, planning, strategising, modelling and implementing solutions to reduce plastic waste, advance plastic circularity and mitigate environmental pollution. The information is also very important for those wanting to set up plastic recycling businesses. At the national level, data provides different values and benefits to the different stakeholders operating in the system, which include the government, interest groups, polymer producers, plastic product manufacturers, consumers, informal waste pickers, waste collectors and managers, sorters and recyclers.

Data helps the government monitor the adoption and effectiveness of its relevant policies introduced to advance plastic circularity. Mapping of waste provides manufacturers with the knowledge of where their waste ends up. In plastic production and consumption, the quantity and type of plastics produced and distributed into the market need to be measured and monitored. For plastic waste managers, defining the quantity, quality and share and supporting the traceability of plastic waste generated, collected and recycled are important.

Plastic Circularity. DOI: https://doi.org/10.1016/B978-0-443-40667-6.00001-9

Improving plastic waste management requires monitoring of the material from the point of production to recycling or safe disposal of the waste, and this requires the implementation of a plastic data collection culture locally, nationally and internationally. Furthermore, data enables the interest groups and environmental organisations to hold all stakeholders accountable for inaction or ineffective implementation.

Furthermore, data is vital in the application of methodologies used for the circularity assessment of a system such as material flow analysis (MFA) and life cycle assessment (LCA). Plastic data needs to be made available to monitor the progress and performance of the transition to circularity in the plastic system. MFA is a methodology commonly used to systematically and quantitatively assess and model the flow of materials within a well-defined system in space and time. LCA can be used to assess the impact of plastic on the environment over its entire lifetime, from raw material extraction to waste management. Having a clearer view of plastic material flows and plastic waste in a plastic system, both at national and local levels, helps to understand some of the challenges facing the system and then plan better for circularity. The effective adoption and integration of the circular economy in the system depends heavily on the availability of data (Çetin et al., 2023), as well as the analysis of the material flowing along the value chain and throughout the life cycle of the material (Davari et al., 2023). Table 5.1 illustrates the importance of data for these circularity assessment methodologies for plastics.

TABLE 5.1 Data impact on circularity assessment methodologies.

Methodology	Importance	Consequences
MFA	• Identifies inefficiencies in the plastic system • Monitors and reduces plastic waste • Improves resource use efficiency • Tracks sources of plastic waste	• Increased plastic waste • Poor plastic waste management
LCA	• Determines detailed impact of plastic • Identifies areas requiring improvements • Improves compliance of products with regulations • Increases reliability of footprint results	• Produces uncertainty in decision-making • Affects reliability of results • Reduces consumers' trust in products

LCA, Life cycle assessment; *MFA*, material flow analysis

Harnessing plastic data to advance circularity and sustainability requires commitments to effective and efficient data collection and reporting on plastics. Regrettably, data collection and reporting are currently a challenge for the global plastics industry, particularly in developing economies. Quantifying and reporting on the production and consumption of plastics, the management of the corresponding waste, as well as the resultant environmental impacts, are increasingly time-consuming, labour-intensive and costly.

5.2 Data sourcing and triangulation

The quantification of the plastics flowing through a typical system usually involves the collection of data from multiple sources, with estimations, extrapolation and mass balancing done in most instances. Plastic data collection is the systematic gathering of quantitative and qualitative data across the plastics value chain for the purposes of evaluation, planning, monitoring and learning. The data collected could either be primary from surveys or estimates from secondary sources. Mostly, the sourcing of plastics data usually involves an extensive search of relevant websites, reports and releases belonging to the government. This is often followed by searches in scientific and non-scientific publications, where relevant data is identified, collated, arranged according to the flow process type or value chain stage and triangulated. The production and consumption of plastics and management of the subsequent waste are sequential processes within the value chain, and this sequence influences the estimation and modelling of data for each of the value chain stages.

Similarly, when plastics are distributed or placed in the market, some are in use or in stock and do not get disposed of immediately or in the same year of consumption. This means it is necessary to also analyse in-use or in-stock plastics. This flow process can be the least quantified, considering that the flow of plastics is usually analysed annually, and as such, data relating to plastics stocks in use and from previous years is not adequately captured. Therefore data and information on plastic product consumption and their associated lifetimes need to be estimated. The end-of-life or waste management stage of the plastics value chain is the most challenging with respect to data availability and estimation. This is as a result of mismanagement practices and activities such as indiscriminate dumping, open burning, littering, leakages and marine pollution. Data on the characteristics of plastic waste is equally lacking in most instances, as disaggregated data on waste is not easily accessible.

In addition, plastic data is sourced from input–output tables, which represent the flow of goods and waste in a system. These tables, which are based on empirical data, have been applied to quantify mass flow of plastics in some countries (Joosten et al., 2000; Nakatani et al., 2020). This approach uses the supply and use tables of a particular country as a source of data in determining the mass flow of a material. According to the tables, the economy of that country is defined in terms of products and services, supply and use of materials, service providers, production industries and consumers (Joosten et al., 1999). The extracted data from these tables is applied in computing the consumption or use of materials and products in the final analysis of the consumption section. The results produced from this approach depend on how detailed the supply and use tables are.

Countries have varying capacities for plastic data sourcing and collection. Depending on the level of development, some countries, particularly developed economies, exhibit proper and sufficient structure and infrastructure and maintain robust data on plastics flow management, including production, consumption, waste generation and management. Other countries, mostly developing economies, lack such structural capacity for data collation on plastics. The availability of plastics data is largely lacking in these countries as they do not have the required technical, human and financial capacity. Efforts are required in these instances to build the know-how, human resources and technical infrastructure in order to improve plastic data generation and collection capacity.

Furthermore, considering that plastic data availability is limited generally in most countries and that relevant data available for the different stages of the value chain is collected from different sources, unifying the available data from the different sources for a specific flow process or value chain stage could be challenging. The types of plastic, the data collection period and the system boundary or geographical area covered have to be unified. Unifying this available data for a specific flow process will often require the data to be triangulated. The triangulation involves the tabulation of all the available data for a process and then cross-examining it to understand the various dimensions of the data, including its alignment. This ensures that all available data collected for a particular flow process is related and credible as far as possible. Having more data collected and more data sources enhances the triangulation process.

5.3 Challenges to plastic data collection

The collection of relevant and reliable plastic data across the value chain comes with many challenges. These challenges include diverse data collection systems and frameworks, variation in data scope, data confidentiality and the lack of incentives to collect and report data on plastics across the value chain.

5.3.1 Diverse data collection systems and frameworks

The different data collection systems, methodologies, strategies and frameworks existing at the provincial, state and municipal levels across countries can be a challenge in the efficient and effective collection of plastic data. The lack of harmonised and mutual frameworks and standards across different jurisdictions within countries has the potential to create uncoordinated collection and reporting of data, in addition to being an administrative burden (Duguay, 2023).

Also, most governance structures for plastic data collection are not centralised but distributed across provinces, states and municipalities. This leads to fragmented collection of plastic data and a lack of adequate national coordination for collection. Having a unified framework for the collection of plastic data can assist in data collection.

5.3.2 Variation in data scope

The lack of a harmonised definition of the scope or system boundary for the collection of plastic data can be another challenge to effective data collection. This relates to the categorisation of different plastic data and information to collect and report, such as plastic polymer types, plastic product applications and industry sectors, among other classifications. Defining the scope uniformly, or mandatory reporting of scope with respect to plastic data collection, will create a complete and broad approach for extensive analysis, comparability and benchmarking of a plastic system. This is important for different actors and operators within the system for informed decision-making.

5.3.3 Data confidentiality

The sharing of plastic data across the plastic value chain can be a major challenge to improved data collection due to data confidentiality. Measures such as the anonymisation of data might be inadequate in resolving this concern. In some instances, the commercial interests of organisations, such as manufacturers and recyclers, and their need to protect their competitive position and customer base, can result in proprietary and fragmented approaches in collecting material flow data. This hinders productive collaboration and data sharing across the value chain.

5.3.4 Lack of incentives

There is often a lack of incentives, interest and efforts by related corporate sectors in the industry to collect and report plastic data, even though they seek, as others do in the industry, to mitigate or reduce the negative impacts of plastics on the environment. The voluntary collection of plastic data is ineffective, as the reduction of the impact of plastics on the environment may not rank top on the priority lists of some businesses. Consequently, the lack of regulations and/or incentives limits the drive for the desired change regarding plastic data collection.

5.4 Data quality and traceability

The variations and complexities in the practices and methods used in collecting plastic data by different entities within a system could result in a varied range of data quality. This is heightened by the absence of mutual plastic data structures, with most data being collected, aggregated and documented manually (Duguay, 2023).

Data traceability of plastic products or items within the system is also an issue. This relates to the process where data across the plastics value chain is tracked with respect to its accessibility, values and changes in order to enable the auditing, verification and validation of the data. While the traceability of plastic data throughout the value chain might be challenging, it is crucial in making the data more credible, particularly with respect to the applications of plastics such as residential, commercial and industrial, as well as features including ethical and proper sources of plastic materials and post-consumer recycled material sources and composition, among others (Duguay, 2023). This is because the various sources for the collection of plastic data have different origins and qualities.

Considering the uncertainty in data due to the data source and the various challenges discussed, it is important to evaluate the reliability and robustness of data collected across the plastics value chain. Evaluating the reliability and robustness of the data can involve either the approach of qualitative estimation or quantitative characterisation of the uncertainty of the data.

The qualitative estimation approach (Galafassi et al., 2019; Pivnenko et al., 2019) is centred primarily on the sources used for data collection, with data collected from direct sources (directly from surveys or empirical sources) categorised as having small uncertainty; indirect sources (from proxy sources such as monetary equivalency and subsequently converted to the required form, e.g. mass value) regarded as medium uncertainty; estimation (projections or estimates made based on empirical data) regarded as large uncertainty and mass balance (data computed by balancing flow inputs and outputs) classified as extra-large uncertainty.

The quantitative characterisation approach is systematically conducted on the data to describe the quality of the plastic data, identify the data gaps and validate the findings or results (Laner et al., 2016). This method consists of two steps and five quality indicators for assessing data, derived from the combination of two assessment concepts, namely, quality assessment scheme (Weidema and Wesnæs, 1996) and the data classification of the material flow data uncertainty (Hedbrant and Sörme, 2001).

For the qualitative assessment aspect of the method, the five quality indicators for assessing the data include reliability, completeness, temporal correlation, geographical correlation and other correlation (Laner et al., 2016). Each quality indicator has a range of scores, descending from good (1) to poor (4). The best data quality is enabled by the following:

Rigour

1. The methodology of data generation is well documented, consistent and based on peer-reviewed data.
2. Value includes all relevant processes/flows in question.

Scope/system boundary

1. Value relates to the right time period.
2. Value relates to the studied region.
3. Value relates to the same product, the same technology, etc.

The indicators of reliability evaluate the documentation of the data generation, whereas those of completeness assess the inclusion of every relevant flow mass in the data point. Temporal and geographical correlation assess the data-point congruence of the analysed system in terms of deviations in time and space, respectively, and other correlation assesses deviations as a result of other factors. After the qualitative data assessment has been completed, the quantitative uncertainties related to the indicators are subsequently assigned, although there is some degree of arbitrariness in this step, particularly where empirical data is unavailable to validate estimates. Normal distribution can be assumed for the data and the uncertainties quantified by coefficients of variation, based on an exponential function, where there is an exponential increase of uncertainty with worse scores.

The coefficients of variation comprised a range of sensitivity levels from 4 to 1, in descending order, with sensitivity scores (for high, medium and low) allocated across the different sensitivity levels for the respective five data quality indicators (Laner et al., 2016). The quantitative uncertainties are computed from this table of coefficients after the score for each indicator is assigned.

5.5 Building data collection capacity

Overcoming the current limitations in the collection of varied data across the plastics value chain requires improving the capacities to collect and report relevant quality data with minimal or no uncertainty. The data collection capacity for plastics can be built and developed in phases and sequences, centring on circular economy principles. Building plastic data capacity in the value chain follows a proposed framework, consisting of four data build-up processes, which are sequential and parallel in certain instances. Data on plastic production connects with consumption, and data on plastic waste generation connects with plastic consumption and recycling. Therefore, data on plastic consumption can be built from plastic production, considering that both plastic production and consumption are processes that flow in sequence, while being complementary at the same time. Data relating to plastic waste management and plastic recycling are sequential.

5.5.1 Data build-up on plastic production

Data on plastic production is the most easily accessible compared to the other stages within the value chain, and the capacity of this data collection point can be further strengthened or built up by the collaboration of relevant stakeholders within this value chain stage. Plastic production data can be collated by plastics manufacturers, as they audit and keep records of relevant sets of data on their production. This provides a good starting point or setting for the build-up of plastic data in the value chain. Associations of plastics manufacturers in a system also keep records of production data, which can be useful for quantification of the plastic flow in such a system. Another important source of data collation is international agencies such as the World Bank, the United Nations and the International Monetary Fund, which generally maintain data records on material imports and exports for various countries globally. National governments and statistical agencies in countries across the world can collaborate with these international systems to build data on plastics produced domestically. The Harmonised System (HS) codes for internationally traded (import and export) primary plastics and plastic products include categories HS 3901 to HS 3914 and HS 3915 to HS 3926, respectively. These are standard classification codes that can be used as a reference in building plastic data and information in the system. At this stage it is important to have a unified framework and scope of classifications for reporting plastic material types and plastic product types.

5.5.2 Data build-up on plastic consumption

Plastic consumption is possibly the most challenging stage in respect of the collection of plastic data in the value chain, as data is largely collected indirectly. Consumption must be associated with or tracked to locally produced and imported plastics. Some plastics are stored in use, as not all plastics will be discarded in the same year, depending on usage, which makes the lifetime analysis of plastics very important. The HS codes can also be helpful in tracking the consumption of imported plastics to build data in the system. Building data on plastic consumption can further be done through surveys of households. The pattern of consumption of the types of plastics and plastic products can thus also be determined. Input–output tables are another means of determining the consumption of plastics in a system, as outlined earlier.

5.5.3 Data build-up on plastic waste generation and management

The plastic waste generation phase links to the phases of production and consumption of plastics. Data on plastic waste generation and management can best be tracked and collected by the local authorities, as they are responsible for waste collection, management and disposal. The local authorities are usually in possession of relevant primary data as they oversee waste management and infrastructure. The integration of the informal sector will boost the collection of important and relevant waste data on plastics through waste pickers. Collected plastic waste data by the different municipalities across the country can successively be transmitted and aggregated at state, province and national levels. Estimates can be made by the municipalities where data is unavailable through the combined efforts of all relevant stakeholders at that level. The capacity of the local authorities in plastic data collection is crucial in the system. Data relating to waste generation and collection rate is important in designing for increased capacity in plastic waste reduction, recycling and circularity.

5.5.4 Plastic waste collection data

For plastic waste to be ready for recycling, it needs to be collected. Tracking data on volume and rates of plastic waste collection is important to drive a circular economy. Municipalities have to set up collection systems that minimise contamination of plastic waste and facilitate cost-effective sorting of waste for recycling. In some countries, an approach is taken that co-collection of plastic, metal and glass presents a lower contamination and material-quality risk. Collecting data on such systems is vital in monitoring and driving system performance.

5.5.5 Data build-up on plastic waste recovery and recycling

Data on recycling is important, as plastic waste recycling is an integral part of the circular plastic economy. Data collection in this phase within the value chain is relatively clear-cut, as long as the relevant actors or operators are involved and carried along in the process. The collection of data on plastic recycling can be built through surveys of recycling companies within the industry. International organisations also keep records on plastic waste imported and exported, which can be estimated for the mass of plastic waste recycled.

5.6 Chapter summary

The role of data in the plastics value chain was explored in this chapter, with each of the value chain stages being examined. The various sources of plastic data collection were highlighted, the various challenges and limitations in the collection of plastic data across the value chain were identified, the impact of the various challenges on the quality of the data was discussed and measures for building data collection capacity across the entire plastics value chain were proposed.

Data in the plastics value chain is crucial in assessing, planning, strategising, modelling and implementing solutions to reduce plastic waste and increase plastic circularity. The development of plastic data usually involves the collection of data from multiple sources, with estimations, extrapolation and mass balancing done in some instances. Therefore, the collection of relevant and reliable plastic data across the value chain comes with many challenges and limitations, such as diverse data collection systems and frameworks, variation in data scope, data confidentiality and the lack of incentives to collect and report data on plastics across the value chain. These challenges limit the quality of the data in terms of credibility, reliability and confidence in data being collected in the plastic system. The various limitations and challenges substantially increase the risks associated with data accuracy, create data uncertainty and lower the quality and value of the plastic data.

A framework was proposed for building plastic data capacity in the value chain, which consists of four data build-up processes: data build-up on plastic production; data build-up on plastic consumption; data build-up on plastic waste generation and management; and data build-up on plastic waste recovery and recycling. Data on plastic production connects with consumption; data on plastic waste generation connects with plastic consumption and plastic recycling. Data on plastic consumption can be built from plastic production, considering that both processes flow in sequence, while being complementary at the same time. Data relating to plastic waste management and plastic recycling is paired, but sequential at the same time.

As a result, manufacturers become more knowledgeable about where their plastic waste ends up; the government is better able to monitor the adoption of relevant policies for more circularity of plastics and interest groups are better empowered to hold all stakeholders accountable for actions and inactions. It helps to understand that data collection has a consequent impact on informed business or environmental decision-making and policy interventions. Tackling the existing challenges and limitations of the collection of varied data across the plastics value chain requires improving capacities to collect and report relevant quality data, with minimal or no uncertainty. Having a unified governance system and framework for the collection of plastic data, possibly a national registry for plastics, can assist in addressing the gaps or challenges associated with individual approaches to collection at the provincial, state or municipal level, enabling better coordination of efficient plastic data collection and reporting, and the tracking of data collection performance. Finally, as with all robust scientific reporting, data needs to be triangulated.

References

Berglund-Brown, J., Kedir, F., Riabova, A., Hall, D., 2022. Assessing Circular Information Flow in Industrialized Construction: a framework for evaluating data for circular construction. IOP

Conference Series: Earth and Environmental Science 1078 (1), 012005. Available at https://doi.org/10.1088/1755-1315/1078/1/012005.

Çetin, S., Raghu, D., Honic, M., Straub, A., Gruis, V., 2023. Data requirements and availabilities for material passports: A digitally enabled framework for improving the circularity of existing buildings. Sustainable Production and Consumption 40, 422–437. Available at https://doi.org/10.1016/j.spc.2023.07.011.

Davari, S., Jaberi, M., Yousfi, A., Poirier, E., 2023. A Traceability Framework to Enable Circularity in the Built Environment. Sustainability 15 (10), 8278. Available at https://doi.org/10.3390/su15108278.

Duguay, D., 2023. Improving data collection, reporting and transparency within the plastics packaging value chain. Circular Economy Leadership Canada (CELC) & Canada Plastics Pact.

Galafassi, S., Nizzetto, L., Volta, P., 2019. Plastic sources: A survey across scientific and grey literature for their inventory and relative contribution to microplastics pollution in natural environments, with an emphasis on surface water. Science of the Total Environment 693, 133499. Available at https://doi.org/10.1016/j.scitotenv.2019.07.305.

Hedbrant, J., Sörme, L., 2001. Data vagueness and uncertainties in urban heavy-metal data collection. Water, Air and Soil Pollution: Focus 1 (3/4), 43–53. Available at https://doi.org/10.1023/A:1017591718463.

Joosten, Hekkert, Worrell, 2000. Assessment of the plastic flows in The Netherlands using STREAMS. Resources, Conservation and Recycling 30 (2), 135–161. Available at https://doi.org/10.1016/s0921-3449(00)00055-0.

Joosten, L.A.J., Hekkert, M.P., Worrell, E., Turkenburg, W.C., 1999. STREAMS: a new method for analysing material flows through society. Resources, Conservation and Recycling 27 (3), 249–266. Available at https://doi.org/10.1016/s0921-3449(99)00009-9.

Laner, D., Feketitsch, J., Rechberger, H., Fellner, J., 2016. A novel approach to characterize data uncertainty in material flow analysis and its application to plastics flows in Austria. Journal of Industrial Ecology 20 (5), 1050–1063. Available at https://doi.org/10.1111/jiec.12326.

Nakatani, J., Maruyama, T., Moriguchi, Y., 2020. Revealing the intersectoral material flow of plastic containers and packaging in Japan. Proceedings of the National Academy of Sciences of the United States of America 117 (33), 19844–19853. Available at https://doi.org/10.1073/pnas.2001379117.

Pivnenko, K., Damgaard, A., Astrup, T.F., 2019. Preliminary assessment of plastic material flows in Denmark: Technical report (Environmental Project No. 2090). Danish Environmental Protection Agency.

Weidema, B.P., Wesnæs, M.S., 1996. Data quality management for life cycle inventories-an example of using data quality indicators. Journal of Cleaner Production 4 (3-4), 167–174. Available at https://doi.org/10.1016/S0959-6526(96)00043-1.

Chapter 6

Modelling of plastic material flows for sustainable decision-making

6.1 Goal and scope definition

Comprehensive knowledge of the material flow of plastics within a country is fundamental in achieving plastic circularity across the plastics value chain system, as well as supporting a circular economy. Material flow analysis (MFA) is an analytical method commonly used to systematically and quantitatively assess and model the flow of materials within a well-defined system in space and time. Expressed in kg/year or kg/capita/year, material flow modelling quantifies and connects the flows between processes, sources and destinations. The principle driving the system is the law of mass conservation, where the mass of a system, closed to every form of transfer of matter and energy, must be held constant over a defined period. Material transfer can only take place from one specific state to another.

Analysing MFA with respect to plastics requires the mass of the plastic material entering a process to be equal to the mass going out of the process, as described by Eq. (6.1) (Kuczenski and Geyer, 2010):

$$\sum \text{Production}_m + \sum \text{Import}_m = \sum \text{Consumption}_m + \sum \text{Export}_m + \sum \text{Stock}_m \tag{6.1}$$

where Production_m, Import_m, Consumption_m, Export_m and Stock_m are the plastic material produced mass, imported mass, consumed mass, exported mass and stock mass, respectively, of plastic material into the system.

The mass conservation of subsequent processes can equally be analysed as:

$$\sum \text{Input}_m + \sum \text{Generation}_m = \sum \text{Product}_m + \sum \text{Transformation}_m + \sum \text{Waste}_m \tag{6.2}$$

where Generation_m is the material produced from other materials within the process and Transformation_m is the material being consumed to produce other items in the system.

Plastic Circularity. DOI: https://doi.org/10.1016/B978-0-443-40667-6.00009-3

MFA has proven to be a useful tool in several fields, such as material management, waste management, wastewater treatment, water management, agricultural management, food production, process control and life-cycle assessment. It is generally a decision- and policy-making tool. Its usefulness and benefits include the optimisation of resource management, early detection of environmental challenges, environmental impact assessments, risk assessments of materials and products, impact assessment of change in consumption patterns, management of waste and environmental pollution, infrastructural and technological development for wastewater treatment, water and sanitation planning and administration, identification and mitigation of nutrient losses in agricultural production and management of nutrients in watersheds.

MFA is an activity that requires various types of data. Regrettably, its implementation can be hampered by several challenges that are data-related, thus limiting its accuracy. These challenges include limited collection capacities in some countries, limited data availability in most processes, unavailability of data in some processes, and low reliability and uncertainties of data as a result of the sources of the data. In these instances assumptions and mass balance computations have to be used.

In this chapter, we define the plastic flows, processes, boundaries and the flow model approaches within the plastics value chain system; examine the various plastic footprints across different sectors, trade balances (import and export) of plastics between countries and recycling performances of plastics and explore various policy mechanisms impacting or targeted at plastics across the different stages of the value chain.

6.2 Modelling flows and system boundaries

The MFA model details the mass of plastic flows, at different stages, across the value chain. It describes the characteristic features of the plastics system and demonstrates how the variables within the system are determined by various parameters. The modelling involves successive steps of a process, which generally incorporate the identification and assessment of the material flow, material stocks, flow processes and the material system. As described by Brunner and Rechberger (2004), the process of modelling a material flow involves three major steps: system analysis, quantification of the flow mass of the material or product and the graphic representation of the flows. The whole process is iterative, thereby enabling repeated examination, revision and improvement of the results.

6.2.1 System analysis

System analysis constitutes the foundation of a material flow model. It entails determining and defining the various variables of the system, such as the relevant or specific materials and products, the flow processes and system boundaries. This requires a detailed knowledge of the phenomena within each flow process and the factors that influence or drive the material flows. This knowledge can be developed by means of a literature review, surveys and fieldwork or expert know-how of the system and processes.

Generally, the modelling of a national plastic flow involves quantifying or accounting for the mass of plastics and plastic waste flowing in and out of the different sectors or boundaries of the plastic system. The system boundary for plastics includes all the main phases in the life cycle of plastics, from the production of polymers to the end-of-life management of the plastic products. The main boundaries usually analysed include the production, consumption, trading, waste collection and end-of-life management of plastic. Illustrated in Fig. 6.1 are the system boundaries and the different flow processes within the plastic system.

The plastic production boundary incorporates every process that contributes to the manufacture of primary plastics and plastic products in the system. Here, the manufacturing process begins with the raw material being transformed from resin or polymer to intermediate and finished plastic products and plastic-embodying items and appliances. Sources of the polymers include local production (virgin polymer), importation and plastic recycling (recycled plastics).

Plastic consumption borders the diverse sectors of the economy, making use of or applying the plastics manufactured. These sectors generally include packaging, health, transportation, automotive, agriculture, textiles, furniture, electrical and electronic appliances and construction. The various products

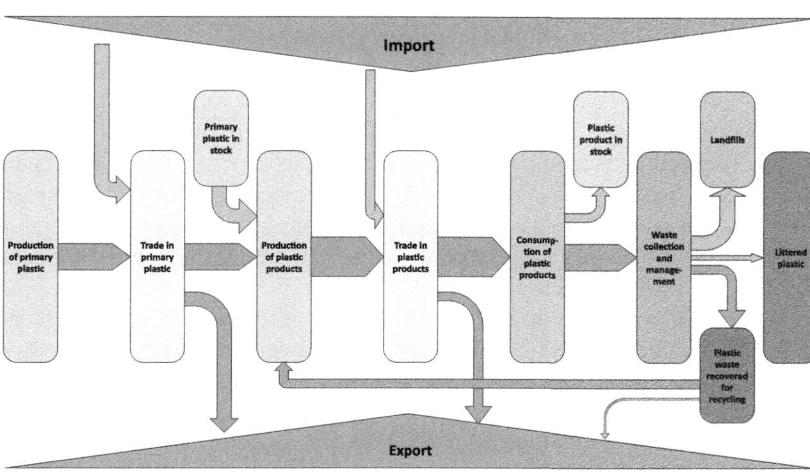

FIGURE 6.1 System boundaries and the various flow processes within the plastic system.

originating from these sectors are largely consumed locally and partially exported. Plastic products for consumption can either be durable or non-durable. Durable plastic products, such as equipment, furniture, electrical and electronic appliances, houseware and motor parts, are associated with long-term application and accumulate in the system. Food packaging, footwear and medical devices are common non-durable plastic products with a shorter lifespan. They are consumed mostly in a single use or over shorter periods.

The import and export boundary of the plastic system constitutes the trading component of the system. The local consumption of both the plastic polymers and plastic products in a country is often, and partly, supported by the process of imports. Equally, some of the plastic polymers and plastic products produced locally are often exported to other countries. Plastic trading surplus occurs where the volume exported is more than that imported, whereas a trading deficit is recorded where the volume imported is more than exportation.

The plastic waste collection and management boundary of the system incorporates the collection and management of various plastic waste streams, which are outputs from the plastic products consumed in the system. Plastic waste that goes to landfills, is recovered for recycling, is not disposed of or is littered is generally quantified here. Plastic waste imported and exported for recycling is also analysed. Local authorities and private waste contractors are generally responsible for the collection of this waste from households, communities, organisations and industries, as well as for the volume collected. The informal sector or waste pickers are an integral part of this boundary in most developing economies globally.

Furthermore, the analysis of the plastic system and the flows can be polymer-specific or sector-specific, which is regarded as a subset of the system. The sector-specific MFA system analysis requires the definition of the assessed sectors and system boundaries. This addresses related sectors such as packaging, healthcare, agriculture, textiles and clothing, transportation, automotive, furniture, electrical and electronic appliances, construction and others. The analysis includes the direct collection of sectoral data from documented records or generation through a transfer coefficient matrix based on literature and assumptions, as well as the computation of the plastic flows along that sector of the plastic system.

The modelling of polymer-specific MFA (polyethylene terephthalate (PET), high-density polyethylene (HDPE), polyvinyl chloride (PVC), low-density polyethylene (LDPE), polypropylene (PP), polystyrene (PS) and others) is generally tied to sectors, as these various polymers are used for different products. It involves a process of selecting the polymers and sectors to be quantified according to available data and information in the literature, assigning the flow mass per sector to the major polymers contributing to the respective sector-specific value chains (top-down approach), generating the polymer-specific transfer coefficient matrix and determining the end-of-life flow mass of the polymers. The top-down approach for modelling polymers entails detailing every sector-specific plastic demand at the level of polymer demand. This is done using demand share, where a polymer-specific share signifies the percentage quantity of the plastic polymer demand in a particular sector. The polymer-specific shares are then computed for the different plastic sectors. Polymer-specific MFA is essential for human and environmental risk exposure assessment, as toxicity differs in the various polymers.

6.2.2 Flow mass quantification

The flows represent the entry of the material or product into the system, the movement from one particular process to another within the system and the final exit of the material or product from the system. Quantification of the flow mass of plastics describes the measurement of the amount of plastics and plastic waste, in kilograms or tonnes, flowing across the system boundaries, from production to the end-of-life management. It includes the record keeping of the mass of all the flow processes that occur within the plastic system, which can be derived either through direct measurement, estimation, projection, extrapolation or mass balance. Data for the quantification of the flow mass is collected mainly through literature or document review.

An important concern in this step is data gaps, as well as the quality (certainty and reliability) of the available data for each flow mass and the data sources. This is a major challenge for many countries across the world, although the situation is often much worse in developing economies. Depending on how the data is sourced, the quality of the data can be classified as high, medium or low. Other forms of classification are small uncertainty (direct measurement), medium uncertainty (indirect source such as conversion from monetary value to mass), large uncertainty (estimations and projections) and extra-large uncertainty (principle of mass balance).

6.2.3 Graphic representation

The final step in the process of modelling a plastic material flow is the graphic representation of the model and flows. This describes the visualisation of the plastic flows and involves the presentation of the results of the flow analysis through diagrams such as pie charts or Sankey diagrams. Represented in Fig. 6.2, the Sankey diagram is the most used flow diagram for material flow representation, where nodes and arrows/links depict the mass and direction of flows between different entities or boundaries within a system.

The width or thickness of the arrows or links is proportional to the flow quantities of the plastics or plastic waste running through the system. The nodes are entities, representing flow processes, actions or process outputs, where flow data enters or exits the plastic system. They are usually illustrated by a box shape and are connected by links. For each node, the input flow

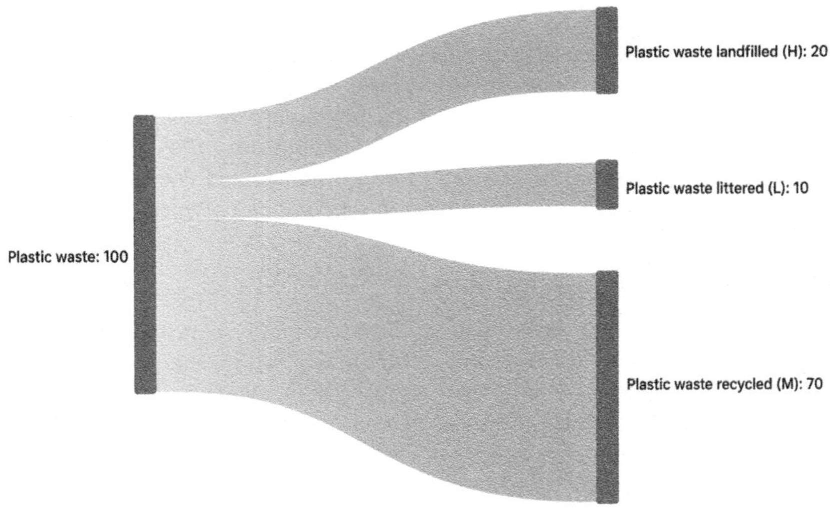

FIGURE 6.2 Sample of a Sankey diagram showing nodes and links. (H) High-quality data; (M) medium-quality data; (L) low-quality data.

equals the output flow to preserve the mass balance of the system. The Sankey diagram is mostly enhanced by colour coding to differentiate between entities and pathways within the system.

6.3 Plastic footprint, trade balance and recycling rate

Knowledge about the national plastic footprint, the plastic trade balance with other countries in terms of import and export, as well as the management of the plastic waste generated, in particular, is important. From the production of plastics through to consumption and end-of-life, plastics interrelate with the environment and the economy in multilayered ways. Import and export also play significant roles in plastic production, consumption and waste management, as international trade occurs in each phase of the life cycle of plastics. Therefore understanding these multilayers is crucial for identifying the various challenges confronting the plastics value chain and formulating an effective policy framework.

Distribution patterns need to be established for primary plastics, plastic products and plastic waste in exploring the footprint, trade balance and recycling rate of plastics. Plastic monomers and polymers are produced from fossil fuel, and depend on the fossil fuel supply of countries. At this stage of production, plastic trade occurs to balance the plastics supply and demand. The products manufactured from plastics are made from polymers, and this process depends on the industrial production capacity of countries. In addition, at this stage, the plastic trade can be used to balance plastic product supply and demand. Plastic waste is generated from the consumption of manufactured plastic products, and this stage of the value chain depends on the consumption capacity of countries. A share of this plastic waste is managed locally, or mismanaged in some instances, as the case may be. The management depends mostly on the ability and capacity of countries to collect and treat plastic waste effectively, particularly through recycling. Another portion of the plastic waste is managed through plastic waste trade, where related importers either recycle the waste for new production or simply discard it.

Recycling retains value for plastic waste after the end-of-life of plastic products, compared to the general pathways of landfilling, incineration (with/without energy recovery) and mismanagement. However, recycling rates for plastic systems are often disputed due to some inconsistencies. Firstly, there are discrepancies in the definition of the concept, with varying designations of the output measure such as *plastic quantities collected for recycling, quantities sent for reprocessing* and *available quantities for use as secondary plastics as a fraction of the total waste*. Also, assumptions sometimes involve the addition of the weight of commercial waste to that of household waste, where it is collected by local authorities. Reported recycling rates may equally be misleading, as computations may be based on easily recyclable plastic as packaging, with difficult plastics rarely reported.

The service life or lifespan of plastic products is another important consideration that determines the length of time plastics will remain in use and eventually become waste. It is strongly associated with the sector in which it is used, as well as how the products are put to use. The service life of plastics in electrical and electronic appliances, water and wastewater, automotive and construction sectors exceeds a year, thus remaining 'in stock' in the system and then released as waste after several years. The average service life of plastics in the building and construction sector ranges from 30 to 50 years, whereas plastics for packaging are used for a year at the most (Geyer et al., 2017).

6.4 Policy interventions across the value chain

Plastic material flow models provide valuable data and information for relevant stakeholders across the plastics value chain to make informed policy decisions for the advancement of plastic circularity and sustainability of the system. National and regional governments are usually given the core mandate and power to regulate the plastics system. As such, they play a crucial role through policy interventions, at both national and subnational levels, in advancing circularity throughout the life cycle of plastics and the sustainability of the environment.

These various interventions can be defined as general waste management policies that are applicable to plastics and policies that are directed explicitly at plastics management. These policies have increased over the years, and they can be categorised into regulations (ban, prohibition, limitation and suspension), economic (subsidies, levy, tax and cash for returns) and information (awareness, campaign, research and education).

From defining plastic waste as an 'end of pipe' problematic material to be disposed of, policy-making around plastics has been redirected over the years to the promotion of sustainable production and consumption of plastics, as well as the restoration of the end-of-life of plastics. This approach places more emphasis on resource efficiency, waste prevention and reduction and environmental sustainability. It entails solutions that involve sustainable product design, intensive use and reuse and recycling of plastics.

To achieve all of this, policies regarding plastic should be able to support every stage of the value or supply chain systems. This will ultimately result in the circularity and sustainability of the plastic system. However, the different phases in the life cycle of plastics have mostly been assessed individually rather than as a whole. The complete value chain or life cycle of plastic is political, with some stages or phases, particularly end-of-life, garnering more public attention and interest. As plastics flow from production to use and then waste management, equal policy support is required for each stage to maximise circularity across the plastic system.

Interestingly, the regulation of the production and use phase of plastic possibly encounters more resistance than regulating the waste management and recycling phase. Regulating plastic waste management is complex, as it is characterised by challenging practices and activities, including waste separation, collection, transportation, sorting and recycling. This is more obvious in developing economies with underdeveloped or poor waste management systems.

Presented in Table 6.1 are samples of policies targeted at the different stages of the plastics value chain system in different countries, which are meant to advance the circularity of those stages.

6.4.1 Deposit refund system

The deposit refund system (DRS) is one of the most potent policy instruments for improving the collection and recycling rates of plastic waste, particularly plastic bottles. The implementation of the policy can impact positively on the recycling stage. Consumers make a deposit of a specific amount when they purchase bottled water or beverages. This amount is incorporated into the price of the water or beverage. These deposits are refunded when consumers return the empty bottles or packaging to a designated location or centre. The policy has the potential to impact consumer behaviour and encourage consumers to be enthusiastic and willing to collect used packaging. As such they are more

TABLE 6.1 Policies and incentives related to the stages of the plastic recycling system.

Stages	Relevant policies	Circularity strategies
Plastic production	• Tax refund in Japan (inclusion of plastic waste in raw material) • Mandatory utilisation in the United States and France	• Tax refund promotes recycling, motivating producers to use plastic recyclate • Mandatory utilisation increases the use of recyclate in production
Product manufacturing	• Extended producer responsibility schemes in some countries in Europe • Tax refund (for infusion of recycled polymer in product manufacturing) • Mandatory utilisation • Plastic bag regulation in China, India, Colombia, Spain, Ghana and South Africa	• Extended producer responsibility increases plastic packaging waste collection for reuse and recycling • Plastic bag regulation reduces the consumption of plastics, thereby encouraging reuse
Waste generation	• Deposit refund system in Norway, Germany and Japan • Kerbside collection system in countries in Europe and America • User pay schemes • Pay-per-bag tax system in Italy	• Deposit refund system promotes the collection of plastic waste for reuse and recycling • Kerbside collection promotes separate waste collection for plastics and thus improves recycling • User pay/pay-per-bag discourages and reduces plastic waste going to landfills
Collection and handling	• Landfill ban in Finland and Austria • Landfill tax in Finland, Sweden and Austria • European Union Landfill Directive	• Landfill ban/tax/directive limits the amount of plastic waste ending up in landfills
Recycling	• Recycling obligation in Norway, Sweden and Denmark • Tax refund (for recycling technology) in Japan	• Recycling obligation promotes and increases plastic waste recycling

enabled and inclined to channel this packaging through the formal collection system rather than the informal system. Similarly, the deposit can influence their purchasing behaviour, thereby limiting consumption. Some European countries, and likewise Japan, have used DRS to standardise the collection of post-consumer plastic bottles after consumption.

6.4.2 Tax refund

The tax refund policy promotes the application of recycled plastics in the production of plastic products. Part of the tax paid by a recycler is refunded when a certain ratio or percentage of the raw material for the production of plastic flakes is sourced from plastic waste. This should lessen the tax pressure on the recycling system, offsetting operational costs in the process. Policy implementation is expected to work as a stimulus for improving plastic recycling, as well as providing adequate renewable resources for the plastic production sector. Countries such as Japan have used the policy to promote the recycling of packaging waste, where different types of recycling equipment for packaging waste are covered, within specific periods, to increase the flow of waste as raw materials into the recycling system.

6.4.3 Mandatory utilisation

This is similar to the tax refund policy as it promotes the use of recycled plastics; however, it is in the form of a charge rather than a refund. Manufacturers of beverage packaging are mandated to use a specific percentage of recyclates in their production to avoid being penalised through fines. This promotes the recycling of plastic packaging, as the recycling sector redoubles its efforts to increase the collection and recycling of packaging waste for the manufacturing sector to meet the prescribed ratios and avoid penalties. However, fines could result in the manufacturers increasing the price of products to retain profit, which could directly affect consumption. France and the United States are examples of countries that have implemented this policy; they have specified a maximum 35% share of recycled materials (Komly et al., 2012) and a planned increase to 25% by 2050 (Tonn et al., 2014), respectively.

6.4.4 Extended producer responsibility

The principle of extended producer responsibility (EPR) requires the manufacturer of a plastic product to take responsibility for the complete life cycle of that product, particularly the end-of-life phase. The policy promotes improved collection of plastic packaging for increased recycling. Manufacturers and importers are often required to pay a prescribed environment fee to have producer responsibility organisations take over the responsibility of managing their products. These are entities legislated or set up by producers to fulfil the packaging recovery and recycling obligations of the manufacturers.

6.4.5 Plastic bag regulation

Plastic bag regulation is used solely to curb the challenge of plastic bag pollution at the production, import, distribution and consumption stages of the value chain. This prohibitive regulatory tool disallows/discourages certain actions relating to the product. For example, plastics could be limited to a maximum quantity or size of plastic material or product permitted at any of these relevant stages; a given type, quantity or size of plastic material or product at any of these relevant stages could be fully or partially prohibited; or the irresponsible handling of some types or sizes of plastic products could be prohibited to address poor waste management practices. National regulations levying or banning plastic bags have increased across countries globally, particularly in Africa, Latin America, the Pacific Islands and the Caribbean (Karasik et al., 2020).

6.4.6 User pay scheme

Also referred to as the pay-as-you-throw system, this scheme is designed to limit the quantity of waste going to landfills by levying a specified fee on households per waste bag for disposal. It could be either a full or partial user pay scheme where a fee is charged for each bag or for bags that exceed the stipulated threshold. The policy provides a disincentive to generate waste for disposal, but promotes the reuse, reduction and recycling of plastic waste.

6.4.7 Landfill tax/ban

The landfill tax and landfill ban were initiated, mostly in some European and Nordic countries, as incentives to limit the mass of waste ending up at the landfill, while promoting incineration and recycling. All waste disposed at landfills is either levied or banned if the utilisation of such waste is technically viable and environmentally friendly. Other waste outside this classification is exempted. The policy is targeted at making waste more commercially utilisable. It has proved to be an effective tool for the diversion of waste from landfill and to improve recycling (European Commission DG ENV, 2012).

6.4.8 Kerbside collection (door-to-door)

Kerbside collection is a waste management system that involves the collection of differentiated household waste, which is placed in separate bags or bins at the kerb or doors of residents. It is a system that supports the separation of different waste streams at source, including plastics from other material waste, and this impacts positively on the quality of waste. This system produces and promotes more returns on the recovery of plastic recyclables in comparison to other waste collection systems such as drop-off and central or communal collection points (Xevgenos et al., 2015; Larsen et al., 2010). This further results in improved recycling of plastic waste. Closeness to collection point and convenience to separate at home drive waste separation at source and recovery of plastics and other recyclables (Zhang and Wen, 2014; Dahlén et al., 2009). This system is most common in the United Kingdom, Sweden, Germany, France and Austria, to mention but a few.

6.5 Chapter summary

Information on and insights into the material flow of plastics in a system are crucial for realising plastic circularity across the plastics value chain system and understanding the trade and environmental risks associated with the use of plastic materials. MFA is an analytical tool to systematically and quantitatively assess and model the flow of materials within a well-defined system in space and time. Material flow modelling is guided by the law of mass conservation, where the mass of the plastic material entering a process must be equal to the mass going out of the process. It is a policy-making tool with numerous benefits, including resource management, resource optimisation, waste management and environmental pollution control.

The development of the plastic MFA model details the mass of plastic flows and their various processes across the value chain. The modelling of national plastic flows involves quantifying the various masses of plastics and plastic waste flowing into and out of the different sectors or boundaries of the plastic system. The process involves three major steps: system analysis, quantification of the flow mass of the material or product and the graphic representation of the flows. The system analysis constitutes the foundation of a material flow model, defining the variables of the system such as the relevant or specific materials and products, the flow processes and system boundaries. The system boundary for plastics includes all the main phases in the life cycle of plastics, from the production of polymers to the end-of-life management of plastic products. Flow quantification of plastics involves the measurement of the mass of plastics and plastic waste flowing across the system boundaries. The graphic representation of the model describes the visualisation of the plastic flows and presentation of the flow analysis results generally through Sankey diagrams.

From production to plastic trade, distribution and end-of-life management, plastics interrelate with the environment and the economy in multilayered ways. Therefore plastic flow models provide valuable data and information to make informed policy decisions for advancing the circularity of the plastic system and the sustainability of the environment. Policy interventions have been made over time by national governments and international agencies and organisations to promote circularity at different phases of the plastics life cycle. These interventions include general waste management policies applicable to plastics and those that are explicitly directed at plastics. They include DRS, tax refunds, mandatory utilisation, EPR, plastic bag regulation, user pay schemes, landfill taxes/bans and kerbside collection systems for waste separation and management.

References

Brunner, P.H., Rechberger, H., 2004. Practical handbook of material flow analysis. The International Journal of Life Cycle Assessment 9 (5), 337–338. Available at https://doi.org/10.1007/bf02979426.

Dahlén, L., Åberg, H., Lagerkvist, A., Berg, P.E.O., 2009. Inconsistent pathways of household waste. Waste Management 29 (6), 1798–1806. Available at https://doi.org/10.1016/j.wasman.2008.12.004.

European Commission DG ENV, 2012. Use of economic instruments and waste management performances. European Commission. Available at https://ec.europa.eu/environment/pdf/waste/final_report_10042012.pdf.

Geyer, R., Jambeck, J.R., Law, K.L., 2017. Production, use, and fate of all plastics ever made. Science Advances 3 (7). Available at https://doi.org/10.1126/sciadv.1700782, https://doi.org/10.1126/sciadv.1700782 Available from.

Karasik, R., Vegh, T., Diana, Z., Bering, J., Caldas, J., Pickle, A., Rittschof, D., Virdin, J., 2020. 20 Years of Government Responses to the Global Plastic Pollution Problem: The Plastics Policy Inventory. NI X 20-05. Duke University, Nicholas Institute for Environmental Policy Solutions, Durham, NC.

Komly, C.-E., Azzaro-Pantel, C., Hubert, A., Pibouleau, L., Archambault, V., 2012. Multiobjective waste management optimization strategy coupling life cycle assessment and genetic algorithms: Application to PET bottles. Resources, Conservation and Recycling 69, 66–81. Available at https://doi.org/10.1016/j.resconrec.2012.08.008.

Kuczenski, B., Geyer, R., 2010. Material flow analysis of polyethylene terephthalate in the US, 1996–2007. Resources, Conservation and Recycling 54 (12), 1161–1169. Available at https://doi.org/10.1016/j.resconrec.2010.03.013.

Larsen, A.W., Merrild, H., Møller, J., Christensen, T.H., 2010. Waste collection systems for recyclables: An environmental and economic assessment for the municipality of Aarhus (Denmark). Waste Management 30 (5), 744–754. Available at https://doi.org/10.1016/j.wasman.2009.10.021.

Tonn, B., Frymier, P.D., Stiefel, D., Skinner, L.S., Suraweera, N., Tuck, R., 2014. Toward an infinitely reusable, recyclable, and renewable industrial ecosystem. Journal of Cleaner Production 66, 392–406. Available at https://doi.org/10.1016/j.jclepro.2013.11.008.

Xevgenos, D., Papadaskalopoulou, C., Panaretou, V., Moustakas, K., Malamis, D., 2015. Success Stories for Recycling of MSW at Municipal Level: A Review. Waste and Biomass Valorization 6 (5), 657–684. Available at https://doi.org/10.1007/s12649-015-9389-9.

Zhang, H., Wen, Z.G., 2014. The consumption and recycling collection system of PET bottles: A case study of Beijing, China. Waste Management 34 (6), 987–998. Available at https://doi.org/10.1016/j.wasman.2013.07.015.

Chapter 7

Participation and dynamics of stakeholders in the value chain

7.1 Value chain coordination

The plastics value chain is a system that represents the flow of plastic material, plastic products and plastic waste. Generally, the end product is the goal of the value chain, with the chain designed around every activity and process needed to produce plastic products. It comprises stages linked together, and each stage in the plastics value chain is expected to contribute or add value to the system, with respect to the set goal. The plastics system is dynamic and multi-staged, with multiple stakeholders in every stage. The linear plastics value chain represents a primary source of accumulated plastic waste and plastic pollution, with a negative impact on the environment. This requires a shift towards a circular plastics value chain where circularity solutions are implemented across the entire value chain to ensure circular product design, production, use and waste management, and not just solely within the end-of-life phase of plastic (Johansen et al., 2022). It further requires the participation of all stakeholders related to each stage of the value chain.

The plastics value chain incorporates all entities or stakeholders participating in the production, distribution, importing, exporting and consumption of plastics, as well as the management of the corresponding waste. Every actor within the value chain pursues a central objective, along with their own individual objectives, which makes participation and relationships in the system complex, thus requiring coordination (Handayati et al., 2015).

Coordination of the value chain describes the integrative participation and interdependence of all the actors based on well-articulated responsibilities and expectations (Duygan et al., 2021). It involves the management of the interdependencies among independent actors within the chain who are all working towards a mutual goal. This is particularly critical with the existence of information asymmetries. Coordination requires the design and development of participatory frameworks for functional collaboration, insightful research, data sharing, information dissemination, efficient resource utilisation and waste resource accountability and traceability. Identifying and addressing

Plastic Circularity. DOI: https://doi.org/10.1016/B978-0-443-40667-6.00010-X

the different challenges confronting various stakeholders operating across the value chain is also important and requires coordination.

The literature has emphasised the importance and benefits of partnership and collaborative efforts in advancing circularity and sustainability practices within the value chain (Del Giudice et al., 2020; Riva et al., 2021). Coordination fosters innovation, efficiency and shared goals among different stakeholders, thereby driving sustainable and circular values. For example, the effective coordination of the relationship between households, waste collectors and plastic recyclers is crucial for promoting circularity principles and processes regarding plastic waste management, such as separation of waste, efficient waste collection, sorting and recycling. The performance of the value chain system improves as each actor contributes unique resources and priorities to the system. Plastic systems with well-coordinated value chains yield many benefits, including a cost-efficient system, increased household recycling awareness and motivation, better waste handling and sorting, improved regulation enforcement, reduced expenses from cost sharing (facilities, infrastructure, machines, equipment) and advancement of inclusive policy mechanisms (Salmenperä, 2021).

Establishing genuine collaboration and effective coordination of the value chain requires the commitment of stakeholders to certain important values or elements such as transparency, disclosure and reporting, as well as communication (Abdelmeguid et al., 2024). These values are essential for enabling stakeholder engagement, building trust and driving circularity practices across the plastics value chain. Transparency emphasises the importance of honesty, openness and accountability in operational practices and processes. Disclosure and reporting ensure that sensitive or personal data and information are shared and made accessible to relevant and interested stakeholders in an understandable manner, thereby enhancing stakeholder engagement. Communication facilitates discussions and information exchange, as all stakeholders have different informational needs. It ensures that every relevant stakeholder is aware of and understands their expected roles and responsibilities, and is supportive of the present and desired future of the system.

7.2 Stakeholders across the value chain

Persons, groups or organisations with a stake or vested interest in the activities, practices, processes or outcomes of the plastics value chain system are described as stakeholders. Their influence on the system can be direct or indirect. Likewise, they can affect or be affected by the system. The support of stakeholders is required for the success of the plastics value chain system.

Stakeholders within the plastics value chain can exercise significant power through conduct that demands action, and this creates responsibility and accountability for the system. This power can either be coercive, where the government uses regulations such as bans and prohibitions to mandate actions, or utilitarian, where protests such as boycotts, blacklisting, or picketing are used by advocacy groups to threaten or disrupt activities or processes of government agencies or private organisations (Phillips and Caldwell, 2005). Stakeholder engagement can reduce friction, promote dialogue and support the effective performance of the plastics system.

The plastics value chain system consists of a diverse range of stakeholders. These stakeholders may be similar across different systems or countries; however, the specific agencies, institutions, organisations, associations, households and businesses in practice may differ in the local context and priorities. The stakeholders within the plastics system can be categorised into internal and external stakeholders (the formal sector), according to the approach by Freeman et al. (2010), and also the informal sector. The internal stakeholders protect or exhibit their interests through direct engagement or involvement in the practices and activities of the plastics system. They include plastic material producers, plastic product manufacturers, distributors and retailers, consumers, waste collection contractors, plastic sorters, buy-back centres (BBCs) and plastic recyclers. The external stakeholders are indirectly linked to the plastics system, but they impact or are impacted by practices within the system. Their interests are exhibited through indirect engagement or involvement in the system. They include government (national, regional and local), research and academic communities, industry associations and environmental or advocacy groups. The waste pickers make up the informal sector, operating at the fringe of the formal sector. They recover plastic recyclables, which contribute significantly to diverting plastic waste from landfills.

The various stakeholders operating within the chain, both formal and informal, constitute a dynamic ecosystem that continues to drive progression in the circularity of the system and the sustainability of the environment. Fig. 7.1

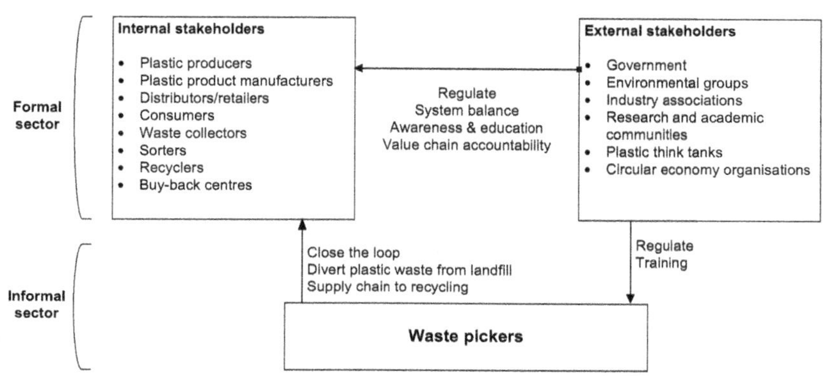

FIGURE 7.1 Stakeholders and their impact and interactions within the plastic value chain. An illustration of the stakeholders in the plastic value chain system.

presents the potential impact and interactions of stakeholders within the plastics value chain.

Each stakeholder contributes to the advancement of the system. The government, whether at the national, regional or local level, plays a chief role in the operation of the value chain, coordinating various relevant activities such as plastic manufacturing and waste management, as well as enabling regulatory and institutional frameworks for the system. They introduce policies and plans, regulate the system and enforce compliance and industry standards. The polymer producers are firms involved in the production of resins for use in the manufacture of plastic products or application in various industrial sectors such as healthcare, packaging, automotive or construction. Plastic product manufacturers are responsible for converting resins into various products such as consumer goods, furniture, packaging and appliances, for consumer use and industrial applications. They can also engage in circularity responsibilities such as the Extended Producer Responsibility (EPR) scheme, which gets them involved in the separate collection of packaging waste for the advancement of recycling. Distributors and retailers represent the bridge between plastic product manufacturers and consumers, showcasing, providing access and selling plastic products to consumers.

Consumers are the users of the manufactured plastic products, and they include households, businesses, industries and public institutions. They exert a sizable influence through their purchasing decisions, thus shaping market preferences and driving demand. They can also be described as generators of plastic waste after consumption. The waste collection contractors are private firms contracted largely by local authorities for the collection and handling of general waste and recyclables. They are sometimes responsible for the sorting and treatment of this waste, as well as the sale to recyclers. Furthermore, the different collection schemes have an impact on the quality of the waste.

The waste pickers represent the informal sector. They generally separate waste and recover post-consumer recyclables, plastics inclusive, from household waste bins and landfills. Found largely in developing and underdeveloped economies, they carry out their operations at the periphery of the formal system and are not registered in the waste management systems of most countries. They play a significant role in the diversion of recyclables from landfills and in the supply of a huge share of plastic waste to recyclers (Growth through innovation annual review, 2016).

The BBCs equally play a significant role in the diversion of plastic recyclables from landfill, acting as the link between the informal waste collectors and the formal recycling firms and manufacturers (Viljoen et al., 2019). As a result, they have attributes of both the formal and informal economy. They are the central recyclable collection point, buying from the waste pickers and selling to the recyclers and manufacturers. They are involved in the sorting, cleaning and baling of the bought waste or recyclables. Plastic recyclers are responsible for recycling plastic waste. They recover plastic waste and reprocess it into other useful products or resins. These new products can sometimes be in forms that are different from the initial form.

The industry associations build relationships between intra- and inter-stages of the plastics value chain. They coordinate data and information within and outside their value chain stages, provide capacity building, training and technical assistance and advance advocacy and policy education. There are various associations for polymer producers, product manufacturers, plastic recyclers and other actors within the different stages of the value chain. The environmental or advocacy groups generally promote and advocate for circular economy practices across the plastics value chain. They hold the value chain accountable and are also central for effective awareness, informational campaigns, education and advocacy within and outside the plastics system.

7.3 Participation in higher waste management hierarchy options and circularity

As the plastics system shifts from the linear model to circularity, the effective participation of stakeholders in the value chain becomes vital for successful implementation. Their roles in advancing higher waste management hierarchy options and the circular economy for plastics are pivotal. The complexities of integrating circularity in the system can be navigated by stakeholders building purposeful relationships, allowing diverse views and promoting collaboration. In this way, the diverse needs, interests, aspirations and expectations of the different stakeholders in achieving circularity success are deliberated and aligned. Consequently, collective knowledge is harnessed, valuable information and insights are gained, "ownership" of the system is built and, ultimately, acceptance of and support for plastic circularity initiatives and practices are increased.

Stakeholders are expected to collaborate in the development and implementation of circularity strategies, policies and practices, which involve innovating and designing for material reduction, intense use, reuse, repurposing and recycling. Collaboration among all the diverse stakeholders across the plastics value chain enables a shared knowledge and understanding of the circular economy vision and goals, thereby facilitating coordinated value chain strategies and actions. The contribution of diverse viewpoints from the various stakeholders should result in innovative circularity solutions for the advancement and sustainability of the system.

Furthermore, it is important to ensure that hindrances or barriers to collaborating or engaging stakeholders are addressed. The diversity of the stakeholders can constitute a limitation. Navigating this and other potential barriers will involve putting in place effective measures. One such measure is the identification and assessment of stakeholders, where potentially relevant stakeholders that could influence the adoption of plastic circularity are identified and assessed comprehensively. Their visions, interests, concerns, strengths, weaknesses and scale of the system boundaries within which the stakeholders operate are some of the elements that should be considered for assessment. Very useful insights for successful collaboration may be acquired from this exercise.

Another important measure, after the stakeholders have been engaged for participation, is education and awareness. Information and knowledge about plastic circularity and its principles, benefits, solutions, practices and system goals need to be raised among stakeholders through awareness, advertisements, seminars, workshops, education, training and targeted communication. Resources, case studies and plastic data can be provided to aid understanding. Concerns raised by stakeholders must also be addressed through evidence-based data, information and transparency. This measure helps alleviate fear and build support for the system goals.

The provision of support mechanisms such as financial incentives, subsidies, grants and resources is another measure that enables active participation of stakeholders. These support mechanisms must be unique and tailored to the needs of the stakeholders. Additionally, mechanisms for monitoring and evaluating the value chain system should be established and continuously practised for the successful integration and implementation of circularity. These should involve the regular assessment of performance and maturity of the value chain stages and system, collation and analysis of insights and concerns of stakeholders and analysis of feedback. The process should also be iterative to enable timely resolutions and adjustments for effective and total circularity of the plastic system.

In a circular plastics value chain model, various stakeholders contribute to the integration of circularity principles and practices in the flow of plastics across the value chain through certain distinct responsibilities and actions. These required responsibilities towards plastic circularity, with respect to each stage of the plastics value chain, are presented in Table 7.1.

7.4 Global partnership towards plastic circularity

Plastic waste is a global problem. Its poor management continues to result in widespread pollution and has a negative impact on the ecosystem, biodiversity and human health. Measures (such as prevention, reduction, reuse, recycling) required to address the plastic waste crisis and mitigate its impacts need to be supported and implemented by all the various stakeholders in the system. Furthermore, these measures need to be promoted, not just at national, regional or local levels, but also at global or international levels.

The essential role that international cooperation is required to play in addressing plastic pollution and increasing the circularity of plastic has been acknowledged in key global agreements and forums. The call for global action in attending to these issues has repeatedly received attention at the international level, including from prestigious international bodies such as the United

TABLE 7.1 Required contributions of stakeholders to each stage of the circular plastics value chain.

Value chain stages	Related stakeholders	Circularity responsibilities and contributions
Plastic material production	• Plastic producers	• Source sustainable and recyclable plastic materials that are reusable or regenerative within the value chain • Include more plastic recyclate than virgin polymers in production • Practise resource efficiency
	• Government	• Introduce a broad plastic policy framework for sustainable product design • Create an enabling environment through incentives and subsidies • Regulate and enforce compliance by the producers • Provide support and funding for research on circular production innovations and technologies
	• Environmental groups and industry associations	• Increase promotion, awareness and education regarding circular plastic production and bio-based plastics • Hold producers accountable for circularity practices • Influence and advocate for a broader plastic policy and institutional framework on circular strategies and practices, stricter regulations and enforcement and total compliance by producers
	• Research and academic communities	• Conduct research and educate on circularity of plastic production (sustainable product design, bio-based plastics, innovations, technologies, infrastructure, policies, practices) • Collaborate with industry in implementing solutions that advance plastic circularity • Provide expertise to producers, governments and policy- and decision-makers

(*Continued*)

TABLE 7.1 (Continued)

Value chain stages	Related stakeholders	Circularity responsibilities and contributions
Plastic product manufacturing	• Plastic product manufacturers	• Develop and implement circular business model • Design plastic products for intense use, reuse, repair, repurposing and recycling • Include plastic recyclates in production • Promote bio-based plastic production • Create opportunities for refurbishment and remanufacturing • Invest in circular economy research and development, innovations and technologies • Participate in producer responsibility or take-back programmes to increase plastic collection for improved recycling • Adopt a closed-loop supply chain through collaboration with raw material suppliers, consumers, waste managers, recyclers and other related entities
	• Government	• Introduce a broad plastic policy framework for sustainable product design and manufacturing • Create an enabling environment through incentives and subsidies • Regulate and enforce compliance by the manufacturers • Enforce mandatory purchase of recycled plastic products by government entities • Provide support and funding for research on circular manufacturing innovations and technologies
	• Distributors/ retailers	• Promote durable and reusable plastic products with longer lifespans • Provide repair services for damaged and repairable plastic products • Participate in take-back schemes or programmes for increased recycling
	• Environmental groups and	• Increase promotion, awareness and education regarding circular plastic product manufacturing

(*Continued*)

TABLE 7.1 (Continued)

Value chain stages	Related stakeholders	Circularity responsibilities and contributions
	industry associations	• Hold manufacturers accountable for circularity practices • Influence and advocate for a broader plastic policy and institutional framework on circular strategies and practices, stricter regulations and enforcement and total compliance by manufacturers
	• Research and academic communities	• Conduct research and educate on circularity of plastic product manufacturing (sustainable product manufacturing, remanufacturing, innovations, technologies, infrastructure, policies, practices) • Provide expertise to manufacturers, governments and policy- and decision-makers
Plastic use and waste generation	• Consumers	• Influence the market towards circularity and sustainability through the demand for plastic products that are eco-friendly, made from recycled plastics, durable or repairable • Prioritise extended use, reuse, repair and repurposing over disposal • Reduce plastic consumption and extend the lifespan of plastic products by adopting, sharing, renting, community gifting and second-hand markets • Participate in take-back programmes • Participate in separation of waste at source and recovery of plastic recyclables for recycling • Become knowledgeable and informed about the environmental impact and sustainability of plastic products • Create awareness within groups or neighbourhoods
	• Waste management associations	• Liaise and build relationships with households, waste collectors and other relevant stakeholders

(Continued)

TABLE 7.1 (Continued)

Value chain stages	Related stakeholders	Circularity responsibilities and contributions
		• Disseminate information and create awareness of separation of plastic waste and recycling
	• Research and academic communities	• Conduct research and educate on sustainable plastic consumption, reuse, repair, repurposing and waste separation
	• Government	• Introduce incentives or levies • Mobilise, disseminate information and create general awareness of plastic reuse, repair, repurposing, waste separation and recycling • Regulate and enforce compliance by households and communities
Plastic waste collection & handling	• Waste collectors and managers	• Promote separate waste collection and sorting for improved plastic waste recycling • Invest in innovations, technologies, infrastructure and human resources for increased and improved plastic waste management and recycling • Partner with governments, households and businesses to develop and implement efficient and effective collection systems and closed-loop plastic material flow
	• Government (municipalities)	• Introduce a plastic policy framework for separate waste collection and plastic recyclable recovery • Regulate and enforce compliance with plastic waste collection • Promote plastic recycling by coordinating the link between waste collectors, sorters and recyclers • Provide and manage infrastructure for improved collection towards increased plastic recycling • Provide training and build capacity through workshops for improved plastic waste collection

(*Continued*)

TABLE 7.1 (Continued)

Value chain stages	Related stakeholders	Circularity responsibilities and contributions
	• Informal collectors	• Regulate operations and activities for improved collection • Provide training to waste pickers for skills and competence acquisition to improve their effectiveness and efficiency in separate waste collection and sorting of plastic recyclables • Provide safety and protection for improved separate waste collection • Prioritise reward system in terms of employment, income and health insurance
	• Research and academic communities	• Conduct research and educate on improved collection, separate waste collection and recovery of plastic recyclables (innovations, technologies, infrastructure, policies, practices) • Provide expertise to collectors and managers, governments and policy- and decision-makers
	• Environmental groups and industry associations	• Increase promotion, awareness and education regarding improved collection and separate waste collection • Hold waste collectors accountable for circularity practices • Influence and advocate for a broader plastic policy and institutional framework on circular strategies and practices, stricter regulations and enforcement and total compliance by waste collectors
	• Waste management associations	• Improve partnership building among waste collectors and between waste collectors and municipalities • Coordinate information and awareness regarding separate waste collection and plastic recycling • Provide training and build capacity for separate waste collection and plastic recycling

(Continued)

TABLE 7.1 (Continued)

Value chain stages	Related stakeholders	Circularity responsibilities and contributions
		• Support the provision and management of infrastructure provision through acquisition, construction, rehabilitation and expansion
Plastic sorting	• Waste collectors/ sorters	• Invest in plastic waste sorting research and development, innovations and technologies • Increase training for human resources • Participate in take-back programmes to increase plastic collection
	• Government	• Introduce policy framework for plastic sorting • Create an enabling environment through incentives and subsidies • Regulate and enforce compliance with plastic waste sorting • Provide support and funding for research on plastic sorting innovations and technologies
	• Environmental groups and industry associations	• Increase promotion, awareness and education regarding improved plastic waste sorting • Hold sorters accountable for circularity practices • Influence and advocate for a broader plastic policy and institutional framework on circular strategies and practices, stricter regulations and enforcement and total compliance by sorters
	• Research and academic communities	• Conduct research on improved plastic waste sorting (innovations, technologies, infrastructure, policies, practices) • Provide expertise to sorters, governments and policy- and decision-makers
Plastic recycling	• Plastic recyclers	• Establish a closed-loop plastic waste flow to reprocess and reintroduce plastic materials effectively into the value chain

(Continued)

TABLE 7.1 (Continued)

Value chain stages	Related stakeholders	Circularity responsibilities and contributions
		• Standardise systems and processes for managing waste streams • Invest in research, technologies, infrastructure and human resources for increased and improved plastic recycling and upcycling • Provide training on plastic recycling and upcycling
	• Government	• Introduce a broad plastic policy framework for improved recycling and upcycling • Create an enabling environment through incentives and subsidies • Regulate and enforce compliance by the plastic recycling industry • Enforce mandatory purchase of recycled plastic products by government entities • Provide support and funding for research on recycling innovations and technologies
	• Research and academic communities	• Conduct research on improved plastic recycling and upcycling (innovations, technologies, infrastructure, policies, practices) • Provide expertise to plastic waste recyclers, governments and policy- and decision-makers
	• Environmental groups and industry associations	• Increase promotion, awareness and education regarding improved plastic waste recycling • Hold recyclers accountable for circularity practices • Influence and advocate for a broader plastic policy and institutional framework on circular strategies and practices, stricter regulations and enforcement and total compliance by plastic recyclers

(Continued)

TABLE 7.1 (Continued)

Value chain stages	Related stakeholders	Circularity responsibilities and contributions
Recyclate market	• Traders	• Provide quality recyclate in the market • Improve sourcing of manufacturers and buyers of recycled plastic products
	• Buyers	• Increase purchases of recycled plastics • Patronise and purchase quality recycled plastics
	• Government	• Create an enabling environment through incentives and subsidies • Regulate and enforce compliance by the market • Enforce mandatory purchase of recycled plastic products by government entities

Nations (UN) and the Organisation for Economic Co-operation and Development. One of these calls targeted at plastics was the resolution adopted by the UN Environment Assembly for setting up an intergovernmental committee by countries to negotiate and develop an international instrument that is legally binding on countries by 2024 with respect to plastics (United Nations Environment Assembly of the United Nations Environment Programme, 2022). This highlighted the role of international cooperation in supporting countries and other entities to address plastic problems and promote circularity. Support from the international community is mostly in the form of either partnering with governments at various levels to reform policies and strategies or partnering and supporting the private sector in promoting increased circularity across the plastics value chain. Additionally, collaborative efforts have been made by two or more countries by combining resources and sharing expertise and experience to advance plastic circularity and sustainability.

With the increasing attention, the number of bilateral and multilateral initiatives extended by the international community to countries to address plastic-related issues is increasing. As such, supportive initiatives continue to increase, and actions regarding the various regulatory, economic and educational support instruments need to be better coordinated. Fig. 7.2 presents samples of these initiatives introduced through international cooperation.

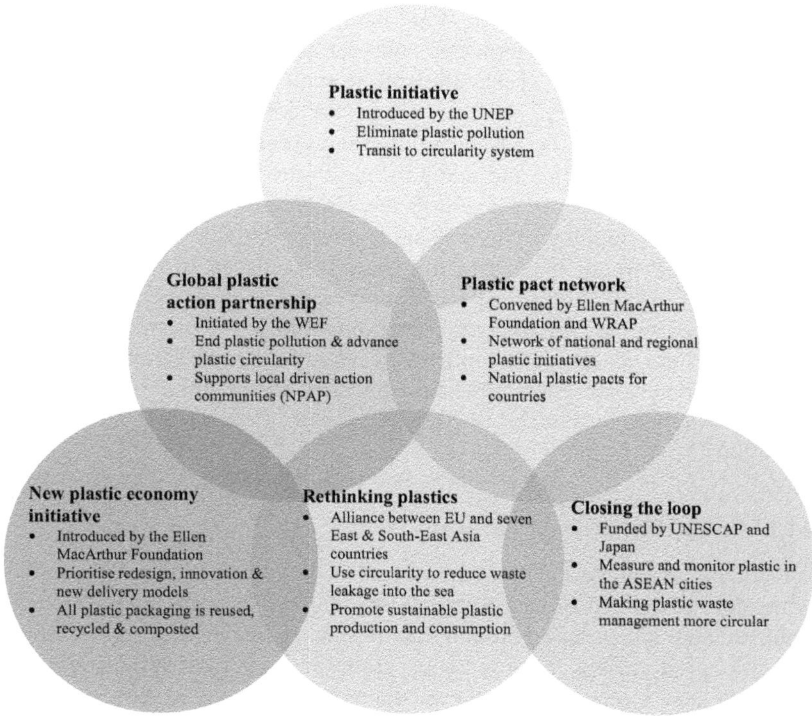

FIGURE 7.2 Plastics-related initiatives through international cooperation.

Some of the globally recognised international entities promoting global partnership and cooperation and introducing initiatives on plastic pollution and circularity of the value chain include the United Nations Environmental Programme (UNEP), World Economic Forum (WEF), Waste and Resources Action Programme (WRAP) and the Ellen MacArthur Foundation.

7.4.1 United Nations Environmental Programme

This global organisation is a foremost authority on matters relating to the environment. Through the United Nations Environment Assembly, it partners with and helps governments, the private sector, advocacy groups and other relevant stakeholders to address environmental problems, including promoting a green and resource-efficient economy. The Plastics Initiative is the platform initiated by UNEP towards ending plastic pollution universally and shifting to a circular system that promotes the design of plastics for reuse, repurpose and recycling. This centres principally on plastic production reduction and increased reuse, rather than sole reliance on recycling.

The initiative is designed on four main goals: reduce the size of the problem through the elimination and substitution of problematic or unnecessary plastics, the market acceleration of reusable plastic products and the increased use of recycled plastics; design for circularity by ensuring that plastic product design is based on the circularity principle of resource efficiency, reuse, recycling and composting; ensure circularity in practice by closing the plastic loop and enabling the circulation of plastic products in practice; and deal with legacy by managing plastic waste that is not reusable or recyclable in environmentally friendly ways. Targets set for the achievement of these goals by 2027 include the improvement of plastics policies in 45 countries towards transitioning to plastic circularity; adoption of circularity solutions to limit plastic pollution by 500 actors in the private sector; and the inclusion of policies promoting plastic circularity and pollution prevention and reduction in the sustainability targets and/or strategy of 50 financial Institutions.

7.4.2 World Economic Forum

The WEF is an international organisation that provides an independent global forum for public-private cooperation on shaping agendas across regions, countries and industries. Through the forum, political, industry, business, academic, civil society, environmental group and other society leaders, with diverse beliefs and views, engage in dialogue and find common ground on global challenges and positive change. The Global Plastic Action Partnership (GPAP) is a foremost initiative of the organisation for accelerating global response to plastic pollution, particularly through the advancement of the circular economy for plastic throughout its life cycle.

Working towards a Global Treaty on Plastic Pollution, the GPAP facilitates and supports the formation of locally owned and locally managed action and impact communities through the National Plastic Action Partnerships (NPAP). Using a multi-stakeholder approach, these national platforms involve the development of national action plans and roadmaps to turn commitments into action with respect to ending plastic waste and pollution and advancing the circular economy for plastics.

7.4.3 Ellen MacArthur Foundation

The Foundation is an international non-profit organisation committed to accelerating the transition to a circular economy, where waste and pollution are eliminated, materials and products are circulated, and nature is regenerated. It targets capacity building for the circular economy, tackling barriers to progress, enabling required conditions and modelling circular economy practices. Partnering with governments, non-governmental entities, policy-makers, businesses, education and academic communities, as well as institutions, it rallies systems solutions globally around the development and promotion of the circular economy. It equally supports systems with research, creates resources and provides learning opportunities.

Regarding plastics, the New Plastics Economy Initiative is the vision of the organisation for the circular economy for plastics. The initiative aims to eliminate unnecessary or problematic plastic packaging by prioritising innovation, redesign and new delivery models; applying reuse models and reducing single-use packaging need; ensuring that all plastic packaging is fully reusable and reused, recyclable and recycled, or compostable and composted in practice; separating plastic use from finite resources consumption; and making plastic packaging free of hazardous chemicals and health and safety related issues.

The Plastics Pact Network is another platform initiated for the circularity of plastics. Co-owned and convened with the WRAP (Waste & Resources Action Programme), the network shares objectives similar to those of the New Plastics Economy Initiative. WRAP is a British-registered charity that works with businesses, individuals and communities to achieve a circular economy. The Plastics Pact Network links regional and national plastic initiatives, bringing together relevant stakeholders to develop and implement geography-specific plastic circularity solutions. Each plastic initiative is administered by a local organisation. The network includes national plastic pacts in the United States, Canada, Chile, Colombia, the United Kingdom, Portugal, France, Poland, South Africa, Kenya and India. The regional pact also includes the ANZPAC Plastic Pact for Australia, New Zealand and the Pacific Islands.

7.4.4 Waste and Resources Action Programme

WRAP is now a global environmental organisation delivering plans, programmes, strategies, research and resources to the benefit of the environment, including reducing waste and restoring nature. The core mandate involves tackling climate change and protecting the planet through changing the system of production, consumption and disposal of materials and products. It brings together various stakeholders, including governments, institutions, businesses and other non-governmental organisations to create a circular and sustainable system that keeps materials in the economy and helps transition from the wasteful linear model to circular and sustainable practices. It supports the use of natural resources in a more sustainable way, promoting reuse and recycling rather than resources becoming waste.

The organisation works with more than 900 organisations globally to end problematic plastics and scale up systemic solutions. It convenes the Plastic Pact Network with the Ellen MacArthur Foundation. The UK Plastics Pact, managed by WRAP, focuses on eradicating needless single-use plastics and increasing plastic packaging recycling and recyclability.

7.5 Chapter summary

The plastics value chain describes the flow of plastic material and waste within a system. This system is dynamic and multi-staged, with multiple stakeholders in every stage. These stakeholders have individual objectives that need to be coordinated. Value chain coordination is the integrative participation and interdependence of all the stakeholders based on well-articulated responsibilities and expectations. It requires the design and development of a participatory framework for functional collaboration, insightful research, data sharing, information dissemination, efficient resource utilisation and waste resource accountability and traceability. Plastic systems with well-coordinated value chains have been shown to yield many benefits, including cost-efficiency, increased household recycling awareness and motivation, better waste handling and sorting, improved regulation enforcement and reduced expenses from cost sharing of facilities and infrastructure.

The plastics value chain system comprises a diverse range of stakeholders. These stakeholders can affect or be affected by the system. Usually, they include the government (national, regional and local), plastic polymer producers, plastic product manufacturers, distributors and retailers, consumers (households, businesses, industries, public institutions), waste collection contractors, the informal sector (waste pickers), BBCs, plastic recyclers, industry associations (manufacturers, recyclers, waste management) and environmental or advocacy groups. They constitute a dynamic ecosystem that continues to drive the progression of the circularity of the system and the sustainability of the environment.

With the plastics system transitioning from linearity to circularity, the effective participation of stakeholders in the value chain becomes crucial for successful implementation. They are required to build purposeful relationships, encourage diverse views and promote collaboration. Achieving all this for the circularity of the plastics value chain will entail important measures that include identifying and assessing stakeholders, providing education and creating awareness, identifying and addressing concerns and financial support mechanisms, and monitoring and evaluating the system. Their role in advancing higher waste management hierarchy options and the circular economy for plastics is fundamental. The various stakeholders are expected to contribute to the integration of circularity principles and practices in the flow of plastics across the value chain through certain distinct responsibilities.

Furthermore, ending plastic waste and advancing plastic circularity are global issues. Circularity measures or strategies required to address the plastic waste crisis and mitigate its impacts need to be supported and promoted at the global or international level. Some of the globally recognised international entities promoting global partnership and cooperation and introducing initiatives on plastic pollution and circularity of the plastics value chain include the UNEP, WEF, Ellen MacArthur Foundation and WRAP. These organisations share a common objective for the advancement of plastic circularity. Some of the plastic-related initiatives introduced through international cooperation include the Plastic Initiative, Global Plastic Action Partnership, Plastic Pact Network, New Plastic Economy Initiative, Rethinking Plastics and Closing the Loop.

References

Abdelmeguid, A., Afy-Shararah, M., Salonitis, K., 2024. Towards circular fashion: Management strategies promoting circular behaviour along the value chain. Sustainable Production and Consumption 48, 143–156. Available at https://doi.org/10.1016/j.spc.2024.05.010.

Del Giudice, M., Chierici, R., Mazzucchelli, A., Fiano, F., 2020. Supply chain management in the era of circular economy: the moderating effect of big data. International Journal of Logistics Management 32 (2), 337–356. Available at https://doi.org/10.1108/ijlm-03-2020-0119.

Duygan, M., Stauffacher, M., Meylan, G., 2021. What constitutes agency? Determinants of actors' influence on formal institutions in Swiss waste management. Technological Forecasting and Social Change 162, 120413. Available at https://doi.org/10.1016/j.techfore.2020.120413.

Freeman, R.E., Harrison, J.S., Wicks, A.C., Parmar, B., de Colle, S., 2010. Stakeholder theory: The state of the art. Cambridge University Press, United States, pp. 1–343. Available at http://doi.org/10.1017/CBO9780511815768.

Growth through innovation annual review, 2016. https://www.plasticsinfo.co.za/wp-content/uploads/2019/09/ebook-2016.pdf

Handayati, Y., Simatupang, T.M., Perdana, T., 2015. Agri-food supply chain coordination: the state-of-the-art and recent developments. Logistics Research 8 (1). Available at https://doi.org/10.1007/s12159-015-0125-4.

Johansen, M.R., Christensen, T.B., Ramos, T.M., Syberg, K., 2022. A review of the plastic value chain from a circular economy perspective. Journal of Environmental Management 302. Available at https://doi.org/10.1016/j.jenvman.2021.113975.

Phillips, R., Caldwell, C.B., 2005. Value Chain Responsibility: A Farewell to Arm's Length* Business and Society Review 110 (4), 345–370. Available at https://doi.org/10.1111/j.0045-3609.2005.00020.x, https://doi.org/10.1111/j.0045-3609.2005.00020.x.

Riva, F., Magrizos, S., Rubel, M.R.B., 2021. Investigating the link between managers' green knowledge and leadership style, and their firms' environmental performance: The mediation role of green creativity. Business Strategy and the Environment 30 (7), 3228–3240. Available at https://doi.org/10.1002/bse.2799.

Salmenperä, H., 2021. Different pathways to a recycling society – comparison of the transitions in Austria, Sweden and Finland. Journal of Cleaner Production 292, 125986. Available at https://doi.org/10.1016/j.jclepro.2021.125986.

United Nations Environment Assembly of the United Nations Environment Programme, 2022. End plastic pollution: Towards an international legally binding instrument. UNEA.

Viljoen, J., Blaauw, D., Schenck, C., 2019. The opportunities and value-adding activities of buy-back centres in South Africa's recycling industry: a value chain analysis. Local Economy 34 (3), 294–315. Available at https://doi.org/10.1177/0269094219851491.

Chapter 8

Progress toward plastic circularity

8.1 The concept of pathways to circularity

Informed decisions and appropriate interventions to advance circularity across the plastics value chain are driven by an approach or process that can identify, compare, assess, estimate and trade off the benefits and impacts of these different circularity strategies (Iacovidou et al., 2017). The pathway to improved plastic circularity defines an approach or process for simulating and mapping out strategies and steps to shift a plastic system from a linear model to circularity, where plastics are designed in a resource-efficient way to last longer, and reuse, repair, repurpose and recycle are prioritised over disposal in landfills. This limits plastic waste going into the environment. The approach is evidence-based and measurable.

Pathways are generally informed by targets set to improve the circularity of the plastic system. As the production of plastics continues to rise globally, with approximately 390.7 million tonnes produced by 2021 (Plasticseurope, 2022), a larger share of this production ends up as waste in landfills. This results in a remorseless accumulation of plastic waste and pollution in the land and water environment. In addressing this problem, countries and institutions are setting targets to limit these tonnes of plastic waste being generated and to make their plastic systems more circular by a certain date. Existing targets largely include cutting down on plastic production, reducing or eliminating plastic waste going to landfills, or increasing plastic waste recycling. Achieving these targets involves the adoption of sustainable and circularity strategies, such as reducing the quantity of plastic waste produced and improving plastic waste recycling. Setting targets is critical for achieving significant progress in the shift towards circularity for plastics, and realising them requires joint efforts by all relevant stakeholders in the system.

A number of benefits can be associated with the modelling of pathways to circularity, including making informed decisions, policy guidance, risk assessment, results-oriented collaboration and time, resource and cost savings (Edirisinghe et al., 2024; Iacovidou et al., 2017). Pathways offer opportunities for stakeholders to make informed and knowledge-supported choices in achieving circularity, as different insights are gained into the potential benefits

Plastic Circularity. **DOI:** https://doi.org/10.1016/B978-0-443-40667-6.00008-1

and limitations of various circularity strategies. They help governments and policy-makers by providing guidance on enacting and introducing knowledge-based circularity plans and policies for plastics.

Potential risks and limitations of certain strategies in achieving plastic circularity can also be detected through pathway modelling before implementation, thus acting as a proactive rather than a reactive measure. Collaborations that are more results-oriented are enabled among relevant stakeholders, as the modelling leads to shared knowledge of potentially effective pathways to transit to improved circularity for plastics. More importantly, selecting the most appropriate pathway for circularity implementation in the system helps to limit plastic waste generation, conserve resources and make the environment more sustainable.

To further define the transition pathways to circular economy objectives and national circularity targets, the generic approach to modelling pathways to circularity is discussed in the next section. Furthermore, considering that there is no one-size-fits-all solution, two main pathways towards improving the circularity of plastics are described. These are the pathways to zero landfill for plastics and the pathways to recycling of plastic waste. These approaches are illustrative and can be adapted to advance other circular plastics economy transitions across the value chain.

8.2 Generic approach to modelling pathways to circularity

Developing an effective solution for improving the circularity of plastics requires knowledge and understanding of the results-oriented potential of different strategies to deliver circularity. To estimate the potential of available strategies under different scenarios, pathways are modelled following fundamental guidelines:

1. Modelling the plastics supply chain stages and their connectivity, defining the various stages of the plastics life cycle or value chain;
2. Analysing current material flows and circularity metrics for the plastics system;
3. Identifying the different available and appropriate circularity strategies to enable circularity, considering the local context;
4. System dynamics (SD) modelling to explore the impact of different interventions and transition actions;
5. Stakeholder engagement and delivery mechanisms;
6. Data analytics.

8.2.1 Modelling the plastics value chain stages

Defining the different stages of the plastics life cycle or value chain categorises the value chain into various phases or stages for the intervention of circularity strategies. Stages of the plastics value chain that are typically considered for circularity interventions include plastic production, plastic use, as well as plastic waste collection, sorting and recycling.

8.2.2 Analysing the material flows of plastics

Material flow analysis (MFA) is a critical component in the shift towards plastic circularity. The analysis of the material flow of plastics provides a quantitative assessment and visual mapping of the movement of plastics and plastic waste across the plastics value chain, with the purpose of optimising resource use and advancing circularity in the system. The mapping is based on the conservation of mass flow from one stage to the next. With respect to modelling pathways, MFA offers vital insights into the consumption patterns of plastics and the potential for plastic reuse and recycling. It helps to optimise the allocation of resources, anticipate disruption to supply and identify inefficiencies and opportunities within the system so as to close loops.

8.2.3 Identifying available and appropriate circularity strategies

Identifying available circularity strategies for intervention at the various stages of the system entails exploring varied waste management options appropriate for the respective stages of the value chain, such as production, use, waste generation, collection, sorting and recycling. Available plastic waste management options across the plastics value chain include reduced production, sustainable product design, bio-based plastics production, reuse, repair, repurpose, separation at source, separate waste collection, take-back programmes, and improved sorting and recycling.

8.2.4 System dynamics modelling

SD represents an analysis of the complex behaviour of the plastic system through feedback from the system (Gao et al., 2020). It analyses the interrelationship and interdependencies of different elements within the plastic system, simulating the interaction of different factors, variables and principles over time within the plastic system in the transition to circularity. Through the model, factors contributing to the challenges to plastic circularity and their impact (cause–effect relationships) are represented through a causal loop diagram (causal connections between different factors), which helps identify the root causes of problems within the system (Dhanshyam and Srivastava, 2021). This helps to address the 'what', 'when' and 'how much' aspects of the circularity interventions when modelling the pathways to circularity. The linearity and non-linearity relationships within the system are incorporated and analysed to obtain tractable interpretations and solutions, making SD more practicable, representative, reliable and adaptable for specific scenarios such as the modelling of the pathways to circularity (Dhanshyam and Srivastava, 2021).

8.2.5 Stakeholder engagement and delivery mechanisms

The actions of stakeholders within the plastic system, with respect to circularity practices, can impact the advancement of circularity across the system. These actions could include deciding on the most impactful circularity strategies to implement and analysing the impact, required processes, investments, returns and costs. Stakeholders' actions in the transition towards plastic circularity can be analysed through agent-based modelling (ABM), as the approach aligns with the concept of the circular economy, and its simulations consider complex and systemic socio-technical interactions and capture the behaviours of individuals or groups (Walzberg et al., 2023). ABM has been applied to related circularity themes such as consumer behaviour, remanufacturing, waste collection and disposal, and waste management and recycling. ABM represents a simulation framework, where agents, depicting individuals, are characterised as unique and independent entities that interrelate with each other and the exogenous environments (Railsback and Grimm, 2019). It estimates complex practical phenomena by studying individuals and their interactions (rather than individual attributes) to capture the resulting behaviour of the system.

8.2.6 Data analytics

Data analytics relates to the application of relevant plastic data and material flows across the value chain to identify opportunities for improved circularity within the system. The process involves the collection, organisation and analysis of data to unearth patterns and correlations within the plastic system. It has significant implications for the plastic system, particularly in promoting circularity across the value chain. Data analytics in the shift towards a circular plastic economy can assist in tracking plastic material and waste within the system, identifying opportunities for recovery of plastic resources, optimising the allocation of resources, improving processes, addressing inefficiencies and enabling data-driven decision-making regarding the development and adoption of circularity strategies (Dhanya et al., 2023). Specifically, vital information and knowledge on plastic material flows, sustainable product design, consumer behaviour and consumption, and plastic waste reduction and management can be acquired by leveraging data analytics.

8.3 Pathway to zero landfill for plastics

The pathways to zero landfill for plastics represent a strategy of diverting waste away from landfills or limiting as much plastic waste as possible from going to the landfills, through waste management policies and circularity strategies and practices. The literature has shown that circularity strategies assist in transitioning the plastic system towards circularity and cut across the plastics value chain (Jiao et al., 2024; Burgess et al., 2021; Du et al., 2013), with some targeting the production stage, including sustainable product design, alternative materials or bio-based plastics, and remanufacture. Some circularity strategies are applicable to the use stage of the value chain, namely, reuse, repair and repurpose. Strategies targeting the end-of-life of plastics include recycling and recovery of plastic waste.

The successful implementation of this pathway will require stakeholder engagement and collaborative actions across the plastics value chain. Crucial actions are required to promote and drive the adoption and implementation of this pathway, particularly through supportive regulatory frameworks and targets. The setting of targets for zero plastic waste to landfill by countries and institutions through various plans and initiatives is critical as it actively drives households, communities and businesses to reduce or eliminate plastic waste generation. This reduces environmental impact, promotes sustainable and efficient use of raw materials, conserves natural resources and saves costs.

As in several other environmental matters, national and regional governments and institutions have introduced initiatives that set targets to achieve zero waste to landfill by a certain date. Some of these target-setting initiatives, national and regional, towards achieving zero landfill include the Landfill Directive (European Union), National Waste Management Strategy (South Africa), Zero Plastic Waste Horizontal Initiative (Canada) and Zero Waste 2040 (Vancouver, Canada).

The European Union (EU), a leading institution in waste matters, initiated the EU Landfill Directive in 2018 (2018). Member States are to ensure a progressive reduction of recyclable or recoverable waste ending up in landfills, as well as introduce measures and processes to limit environmental impact. Specifically, the directive with respect to the zero waste to landfill initiative demands that: *'As of 2030 all waste suitable for recycling or other recovery, in particular in municipal waste, shall not be accepted in a landfill with the exception of waste for which landfilling delivers the best environmental outcome'* and also *'the share of municipal waste landfilled is limited to 10% by 2035'*. This target is based on the principle of landfilling being the least desirable option in the EU waste hierarchy and must therefore be limited to the barest minimum. The main objective of the directive is to improve waste management, generate support for recycling and minimise landfilling.

For South Africa, the Department of Environment, Forestry and Fisheries (DEFF) initiated the zero waste to landfill initiative in 2020 through the National Waste Management Strategy 2020 (DEFF, 2020). Towards achieving this, it specifically set target outcomes for waste management with the mandate to *'Prevent waste, and where waste cannot be prevented ensure – 40% of waste diverted from landfill within 5 years; 55% within 10 years; and at least 70% within 15 years leading to Zero-Waste going to landfill'*. The baseline and expected years for actualisation of policy total outcomes are 2020 and 2050, respectively. Some of the fundamental principles supporting the policy include waste prevention, minimisation and waste as a resource.

Environmental and Climate Change Canada, a federal department responsible for the development and implementation of measures for preventing and reducing the release of plastics into the ecosystem, led other key federal organisations in Canada to launch the Zero Plastic Waste Horizontal Initiative in 2019 (Zero plastic waste, 2024) to achieve zero plastic waste for the country. The initiative is the contribution of the national government to the joint federal, provincial and territorial Canada-wide Strategy on Zero Plastic Waste, requiring multiple organisations to collaborate to achieve the shared objective. It mandates that by 2030, plastics are to be kept in use within the economy and out of the environment. It clarifies that by the reference year, no plastic waste will be sent to landfills and none will be released into the environment. It rather prioritises the reuse, repair, re-manufacture, or recycle of plastics. The chief focus of the initiative is to contribute to the development of the circular economy for plastics in the country, in the belief that nothing is a waste in such an economy.

The Zero Waste 2040 initiative (COV, 2018) was proposed by the City of Vancouver, Canada, in 2016, as a strategic policy for the city to become a zero-waste community and transition to a more circular economy. It connects the city's impact with the waste reduction initiatives of the regions, provinces and nation, as well as globally. Eliminating solid waste from landfills and incinerators by 2040 is the fundamental objective of the initiative. The supporting objectives and approach to actualising the initiative include avoiding, where possible, and reducing waste generation; promoting the reuse of material, including repairing, sharing, refurbishing and repurposing, rather than recycling and disposal; and increasing the recycling and recovery of material. According to the city, the initiative is expected to help address the rising demand for limited waste disposal facilities and support waste-to-resource.

A review of these identified zero-waste-to-landfill initiatives shows that substantially improving the circularity of the plastic system to a level of zero plastic waste to landfill requires varied proposed circularity interventions across the value chain system, which need to be modelled as pathways. These modelled pathways are analyses of various scenarios under which the targets of the different initiatives can be realised.

Therefore according to the reviewed initiatives and the literature, the potential pathway to achieve zero landfill for plastics must be modelled to integrate relevant circularity interventions and practices that centre on reducing or eliminating plastic waste through sustainable plastic product design, substituting plastics with bio-based plastic or alternative material, applying the reuse model (reuse, repair and repurpose) and recycling and recovery. Table 8.1 presents a potential pathway to zero landfill for plastics, which includes circularity strategies, principles and assumptions, as well as the strategic stakeholders.

Reducing or eliminating plastic waste through the sustainable design of plastic products involves innovatively designing out plastic waste while retaining the values and benefits of the material. It requires designing products

TABLE 8.1 Analysis of the pathway to zero landfill for plastics.

Pathway	Target	Strategies	Principles
Zero landfill for plastics	Plastic waste reduction or elimination	Reduce/eliminate plastic waste through sustainable product design	• Significant reduction in the demand for plastic packaging, without compromising on functionality, is technically feasible (Ahmed, 2021). • Practices include reducing secondary plastic packaging and surplus headspace, introducing lightweight models or formats and producing products that are free of packaging and with edible coatings.
		Reduce/eliminate plastic waste through substitution	• Analyses show that substituting plastic with alternatives to improve circularity is highly feasible and beneficial (Singh et al., 2020; Babaremu et al., 2023). • Related practices include using packaging for sauce packets that is fibre-based and compostable.
		Reduce/eliminate plastic waste through a reuse model (reuse, repair, repurpose)	• There should be a continuous rise in the reuse and repair of plastic products. • Adoptions of associated practices include the donation or sales of used plastic products to charity organisations or communities, reusable grocery bags and refill packaging solutions for water or food. • The reuse solutions are environmentally sustainable and technically feasible in considerable proportion, particularly with packaging, according to studies (Coelho et al., 2020).
		Reduce/eliminate plastic waste through increased recycling and recovery	• Studies have shown that there is potential for a substantially expanded recycling system (Salmenperä, 2021). • Supportive plastic recycling policy frameworks (subsidies, incentives and levies), funding, circular business models, waste collection schemes, technology and infrastructure are critical to the success of the pathway.

with less plastic material and rethinking excessive plastic packaging for items, which will help limit a considerable percentage of plastic packaging waste. Substituting plastics with bio-based plastic or other alternative materials entails replacing plastic with materials that are more circular and sustainable. Bio-based plastic is more environmentally friendly and biodegradable, with a limited impact on the environment. Paper and compostable materials are also sustainable alternatives and can disintegrate into natural elements. The reuse model (reuse, repair and repurpose) involves reusing a whole or part of a plastic product after its initial life cycle for successive life cycles. This promotes resource efficiency and prevents waste. It may involve actions such as repair and repurpose. Eliminating or reducing plastic waste through increased recycling and recovery involves improving the reprocessing of plastic waste into new products. It denotes increased plastic recyclate yields and higher quality to reduce or eliminate plastic waste.

8.4 Pathway to recycling of plastic waste

This pathway embodies strategies targeting the end-of-life phase of plastics for improved circularity of the plastic system. The recycling of plastic waste provides potentially valuable and more environmentally friendly feedstock (plastic recyclates). The performance of recycling, in general, in terms of the recyclate quality and ease of recycling, depends on a number of factors (plastic waste type, design, properties, colour, contamination and multilayer or mixed polymer problems), particularly related to the feedstock quality.

Similarly, a number of target-setting plastic recycling initiatives to achieve optimal recycling rates by a certain date have been initiated across countries and continents by governments and organisations. Interestingly, studies have shown that there is potential for a substantially expanded recycling system (Salmenperä, 2021; Milios et al., 2018), which supports the feasibility of any ambitious recycling targets, as well as the pathways to achieve them. Some of these initiatives with recycling targets include the Packaging and Packaging Waste Directive (EU), Resource Circulation Strategy for Plastics (Japan), National Packaging Targets (Australia) and Extended Producer Responsibility Plastic Recycling Targets (India).

Directive 94/62/EC on packaging and packaging waste of the European Commission (European Parliament & Council of the European Union, 2018) was introduced in 2018, and it requires the countries of the EU to take the necessary actions in meeting varying recycling targets with respect to different packaging materials. Particularly, it states that *'by 31 December 2025, at least 65% by weight of all packaging waste must be recycled'* and *'by 31 December 2030, at least 70% by weight of all packaging waste must be recycled'*. The recycling targets for plastic packaging by 2025 and 2030 are 50% and 55%, respectively.

The Resource Circulation Strategy for Plastics was introduced by Japan in 2019 to reduce the consumption of single-use plastics, while promoting substitutes for plastics made from petroleum (MOE, 2019). The initiative specifically requires *'reusable and recyclable design by 2025; reuse/recycle 60% of containers and packaging by 2030; effective use of 100% of used plastics by reuse and recycling etc. by 2035; double the use of recycled content by 2030'*.

Australia commissioned the 2025 National Packaging Targets in 2018 (Department of Climate Change, Energy, the Environment and Water, 2018) to commit the government and the industry to shift towards a circular economy for packaging. Specifically, by 2025, it requires the achievement of *'100% of packaging being reusable, recyclable or compostable; 70% of plastic packaging being recycled or composted; and 50% average recycled content included in packaging'*. The targets are expected to produce a sustainable pathway for the management of packaging in the country.

The Extended Producer Responsibility (EPR) Plastic Recycling Targets (MEFCC, 2022) were initiated in 2022 by India, and it stipulates mandatory targets for producers, brand owners and importers regarding the rigid plastic packaging reuse, the minimum level of plastic packaging waste recycling and the minimum use of recycled plastic content in plastic packaging. The guidelines of the initiative mandate *'30–50% of EPR by 2024-25; 40–60% EPR by 2025-26; 50–70% EPR by 2026-27; and 60–80% EPR by 2027-28 onwards'*, as the plastic waste recycling minimum levels for all categories of plastic packaging. The focus is on reinforcing plastic packaging circularity and recycling infrastructure development for plastic packaging.

An analysis of these plastic recycling initiatives and the literature reveals that the potential pathway to recycling plastic waste requires the recyclability of plastics to be optimised from the product design to the reprocessing stage. As illustrated in Table 8.2, the pathway therefore contains principles and circularity strategies that include innovatively and sustainably designing products for recycling purposes, improving the collection and sorting of plastic waste for recycling and increasing recycling capacity.

Designing plastic products for recycling helps to ensure that when plastics become waste, they can be easily recycled and upcycled, possibly multiple times, while retaining their high qualities and having less impact on the environment. Improving the collection and sorting of plastic waste for recycling involves the improvement of waste separation at source, which increases the quantity of plastic waste recovered for recycling, as well as enabling a higher purity or quality level of the plastic waste stream and, consequently, the recyclate. Increasing the capacity of plastic recycling to support the pathway requires stronger industrialisation, higher standardisation of the sector, improved process optimisation, increased investment, expanded market structures and improved stakeholder engagement and collaboration.

TABLE 8.2 Circularity strategies and practices for the pathway to recycling of plastic waste.

Pathways	Target	Strategies	Principles
Recycling of plastic waste	Optimal recycling	Designing products for recycling purposes	• Adoption of practices such as reducing secondary plastic packaging and surplus headspace, introducing lightweight models or formats and producing products that are free of packaging and with edible coatings
		Improving collection and sorting of plastic waste for recycling	• Maximum plastic waste collection through EPR • Increasing collection of plastic recyclables by informal sector • Increase in separation at source by households and communities • Increase in plastic waste sorting • Improved innovations and technologies for plastic waste sorting technologies
		Increasing recycling capacity	• Potential for a substantially expanded recycling system (Salmenperä, 2021) • Supportive plastic recycling policy frameworks (subsidies, incentives and levies), funding, circular business models, waste collection schemes, technology and infrastructure critical to the success of the pathway

8.5 Chapter summary

This chapter deals with the progress towards circularity of the plastic system. In particular, the concept and significance of pathways to plastic circularity were examined, the generic approach to modelling pathways to circularity was outlined and potential pathways to zero landfill for plastics and recycling of plastic waste were presented.

Retaining the usefulness of plastics while reducing their impact on the environment and biodiversity demands a transition towards a more circular plastics value chain. Appropriate interventions to advance circularity across the plastics value chain are driven by an approach or process that can identify, compare, assess, estimate and trade-off benefits and impacts of these different circularity strategies. This approach or process is termed a 'pathway'.

A pathway to improved plastic circularity is an approach or process of simulating and mapping out various strategies and steps to shift a plastic system from a linear model to circularity, where plastics are designed in a resource-efficient way to last longer, and reuse, repair, repurpose and recycle are prioritised over disposal in landfills. This limits plastic waste going into the environment. The approach is evidence-based and measurable. Benefits related to the modelling of circularity pathways include making informed decisions, policy guidance, risk assessment, results-oriented collaboration and time, resource and cost savings.

Similarly, potential risks and limitations of certain strategies in achieving plastic circularity can be detected through pathway modelling before implementation, thereby acting as a proactive rather than a reactive measure. The selection of the most appropriate pathway for circularity implementation in the system helps to limit plastic waste generation, conserve resources and make the environment more sustainable.

Developing an effective solution for improving the circularity of plastics requires knowledge and understanding of the results-oriented potential of different circularity strategies. To estimate the potential of available strategies under different scenarios, pathways are modelled following fundamental guidelines that include defining the various stages of the plastics life cycle or value chain, identifying the different available circularity solutions or strategies, analysing the material flows of plastics, modelling system and dynamics and producing data analytics.

Furthermore, based on the setting of targets by countries and institutions to improve the circular economy for plastics in their domains, two potential pathways were assessed, namely the pathway to zero landfill for plastics and the pathway to recycling of plastic waste. Informed by related initiatives and the literature, the analysis of the pathway to zero landfill for plastics integrates relevant circularity interventions and practices that centre on reducing or eliminating plastic waste through sustainable plastic product design, substituting plastics with bio-based plastic or alternative material, applying the reuse model (reuse, repair and repurpose) and recycling and recovery. Similarly, the analysis of the pathway to recycling of plastic waste requires the recyclability of plastics to be optimised from product design to reprocessing, comprising strategies such as innovatively and sustainably designing products for recycling purposes, improving collection and sorting of plastic waste for recycling and increasing recycling capacity.

References

Ahmed, A., 2021. Achieving sustainability in the design of Plastic Products. International Design Journal 11 (3), 47–57. Available at https://doi.org/10.21608/idj.2021.162446.

Babaremu, K., Oladijo, O.P., Akinlabi, E., 2023. Biopolymers: A suitable replacement for plastics in product packaging. Advanced Industrial and Engineering Polymer Research 6 (4), 333–340. Available at https://doi.org/10.1016/j.aiepr.2023.01.001.

Burgess, M., Holmes, H., Sharmina, M., Shaver, M.P., 2021. The future of UK plastics recycling: One Bin to Rule Them All. Resources, Conservation and Recycling 164. Available at https://doi.org/10.1016/j.resconrec.2020.105191.

Coelho, P.M., Corona, B., ten Klooster, R., Worrell, E., 2020. Sustainability of reusable packaging–Current situation and trends. Resources, Conservation & Recycling: X 6, 100037. Available at https://doi.org/10.1016/j.rcrx.2020.100037.

COV, 2018. Zero waste 2040. City of Vancouver, RTS 12177.

Department of Climate Change, Energy, the Environment and Water, 2018. 2025 National Packaging Targets. Australian Government. Department of Climate Change, Energy, the Environment and Water. Available at https://www.dcceew.gov.au/environment/protection/waste/packaging/packaging-covenant.

DEFF, 2020. National waste management strategy 2020. Department of Environment, Forestry, and Fisheries.

Dhanshyam, M., Srivastava, S.K., 2021. Effective policy mix for plastic waste mitigation in India using System Dynamics. Resources, Conservation and Recycling 168, 105455. Available at https://doi.org/10.1016/j.resconrec.2021.105455.

Dhanya, D., Kumar, S.S., Thilagavathy, A., Prasad, D.V.S.S.S.V., Boopathi, S., 2023. Data analytics and artificial intelligence in the circular economy. Case studies Intelligent Engineering Applications and Applied Sciences for Sustainability. IGI Global, India, pp. 40–58. Available at https://doi.org/10.4018/979-8-3693-0044-2.ch003.

Du, Y., Cao, H., Chen, X., Wang, B., 2013. Reuse-oriented redesign method of used products based on axiomatic design theory and QFD. Journal of Cleaner Production 39, 79–86. Available at https://doi.org/10.1016/j.jclepro.2012.08.032.

Edirisinghe, L.G.L.M., de Alwis, A.A.P., Prakash, S., Wijayasundara, M., 2024. Optimizing material circularity pathways in industrial waste streams: a decision-making model. MethodsX 13, 102813. Available at https://doi.org/10.1016/j.mex.2024.102813.

EuropeanCommission, Directive (EU) 2018/850 of the European Parliament and of the Council: Amending Directive 1999/31/EC on the landfill of waste. (2018a), Available at https://eur-lex.europa.eu/legal-content/EN/TXT/PDF/?uri=CELEX:32018L0850.

European Parliament & Council of the European Union, 2018. Directive (EU) 2018/852 of the European Parliament and of the Council of 30 May 2018 amending Directive 94/62/EC on packaging and packaging waste. Official Journal of the European Union L 150, 141–154. Available at https://eur-lex.europa.eu/eli/dir/2018/852/oj.

Gao, C., Gao, C., Song, K., Fang, K., 2020. Pathways towards regional circular economy evaluated using material flow analysis and system dynamics. Resources, Conservation and Recycling 154, 104527. Available at https://doi.org/10.1016/j.resconrec.2019.104527.

Iacovidou, E., Millward-Hopkins, J., Busch, J., Purnell, P., Velis, C.A., Hahladakis, J.N., et al., 2017. A pathway to circular economy: Developing a conceptual framework for complex value assessment of resources recovered from waste. Journal of Cleaner Production 168, 1279–1288. Available at https://doi.org/10.1016/j.jclepro.2017.09.002.

Jiao, H., Ali, S.S., Alsharbaty, M.H.M., Elsamahy, T., Abdelkarim, E., Schagerl, M., et al., 2024. A critical review on plastic waste life cycle assessment and management: Challenges, research gaps, and future perspectives. Ecotoxicology and Environmental Safety 271. Available at https://doi.org/10.1016/j.ecoenv.2024.115942.

MEFCC, 2022. 2022 Ministry of Environment, Forest and Climate Change Plastic recycling targets.

Milios, L., Holm Christensen, L., McKinnon, D., Christensen, C., Rasch, M.K., Hallstrøm Eriksen, M., 2018. Plastic recycling in the Nordics: A value chain market analysis. Waste Management 76, 180–189. Available at https://doi.org/10.1016/j.wasman.2018.03.034.

MOE, 2019. Japan's resource circulation strategy for plastics Ministry of the Environment, Government of Japan.

Plasticseurope, 2022. Plastics: The facts 2022. Plastics Europe 11.

Railsback, S.F., Grimm, V., 2019. Agent-Based and Individual-Based Modeling: A Practical Introduction, Second Edition. Princeton University Press, United States, pp. 1–340.

Singh, P., Pirayesh, A.,Mosahebi, A. 2020. Aesthetic surgery during COVID-19. Aesthetic Surgery Journal (NLM Medline), 40, 9, NP566-NP568, https://doi.org/10.1093/asj/sjaa151 https://www.scopus.com/inward/record.uri?eid=2-s2.0-85089606522&doi=10.1093%2fasj %2fsjaa151&partnerID=40&md5=ff5300ffea251ec728833486bac91604.

Salmenperä, H., 2021. Different pathways to a recycling society – Comparison of the transitions in Austria, Sweden and Finland. Journal of Cleaner Production 292, 125986. Available at https://doi.org/10.1016/j.jclepro.2021.125986.

Walzberg, J., Frayret, J.M., Eberle, A.L., Carpenter, A., Heath, G., 2023. Agent-based modeling and simulation for the circular economy: Lessons learned and path forward. Journal of Industrial Ecology 27 (5), 1227–1238. Available at https://doi.org/10.1111/jiec.13423.

Zero plastic waste, 2024. Office of the Auditor General of Canada (OAG). Reports of the Commissioner of the Environment and Sustainable Development. 3.

Chapter 9

Transition towards a world-class system for plastics circularity: Performance and maturity assessment

9.1 Plastic system assessment towards world class

The plastics value chain faces challenges from the proliferation of plastic waste, which continues to result in land-filled waste and pollution. Pressures from government regulatory entities, environmental groups and other interested stakeholders for the value chain to be completely circular continue to grow. In transitioning towards world-class circularity for plastics, it is critical to determine and define the elements and factors that impact or drive the performance and development of the plastics value chain system.

The circularity of the plastic system requires that the pathways to development be improved continuously. The improvement of the pathways therefore requires the performance and maturity of the plastic system to be known and understood. This involves assessing the level of performance and maturity of the whole system, as well as the respective stages of the system. The assessment of the performance and maturity of the plastic system helps in the design of action plans to bridge the gaps between the current level of performance and maturity and the desired optimal performance and full maturity of the system.

9.2 Performance metric framework for transition to a world-class circular economy for plastics

The performance metric framework for the circular economy for plastics provides a set of key metrics for assessing the level of performance of a plastic materials system and benchmarking its success towards optimal circularity performance, regardless of geographical location. The metrics represent critical factors required for increased plastic system performance. The framework is applied to the system to assess the complete value chain and the

Plastic Circularity. DOI: https://doi.org/10.1016/B978-0-443-40667-6.00004-4

respective stages, in order to determine the individual performance of these stages and identify specific ones that require improvement.

The development of the performance framework involved a systematic review of relevant literature to establish the key metrics for performance. The article review identified several factors that drive plastic circularity: subsidies, tax incentives, government finances, household economics, plastic bag limits, kerbside collection schemes, segregation, user pay programmes, public-private cooperation, planning, administration, human resources, skilled manpower, local markets, technology, land availability, general awareness and education (Olatayo et al., 2023).

The key performance metrics, which include policy and incentives, value chain coordination, infrastructure, technical expertise, demand, waste generation scale, recovery value and awareness, were used by Olatayo et al. (2023) to develop a performance metric framework. This framework was particularly plastic recycling-focused and incorporates the key performance metrics, metric objectives for plastic recycling, the relevant stages of the plastics value chain for each performance metric, and the measurement indices to assess the relevant performance metrics.

Following this framework designed for a plastic recycling system by Olatayo et al. (2023), this chapter further modelled a broader performance metric framework for the circular plastic economy, presented in Table 9.1. Applying the key performance metrics, it integrates the circularity targets of the metrics towards advancing plastic circularity, the applicable stages of the plastics value chain for each metric for circularity, and performance indicators to assess the relevant performance metrics. The application of the broader performance metric framework to advance the circular plastic economy involves assessing each stage of the value chain through the stage-related performance metrics and corresponding performance indicators.

The performance metric framework has a principal benefit of helping to guide actions to advance world-class plastic circularity across the value chain.

9.3 Maturity model for plastic system

Following the development of the broader performance metric framework for the circular plastic economy to assess the level of performance of plastic systems towards advancing circularity across different jurisdictions, a maturity model was developed to evaluate the circular economy maturity of a plastic system. The concept of maturity refers to the development or growth of an organisation or system from a preliminary position or status to one that is more advanced (Fraser et al., 2002), incorporating a sequence of stages that represent a desired path to an objective (Poppelbuß and Roglinger, 2011). The integration of these different stages makes up a maturity model, whose principal goal is to enable and promote continuous improvement.

A maturity model is a multi-level, structured and defined path for improvement and progression of a system or organisation (Antunes et al., 2014).

TABLE 9.1 Broader performance metric framework for circularity of plastic systems.

Metric to advance circularity	Metric objectives	Applicable value chain stages	Performance indicators per stage
Improve value chain coordination	• Increase circularity for plastics through improving plastics and plastic waste traceability across the value chain • Improve the availability of plastic data and information quality to improve circularity across the value chain	Plastic production	• Operational cost reduction in circularised sourcing of plastic for production and plastic material reduction • Product differentiation as a result of tracing and circularised sourcing of quality plastic material • More availability of quality plastic production data
		Plastic product manufacturing	• Operational cost reduction in circularised sourcing of produced plastics for product manufacturing and plastic product material reduction • Product differentiation as a result of tracing and circularised of sourcing quality plastics • Increased availability of quality product manufacturing data
		Plastic waste generation	• Increased availability of quality waste generation data • Increased public-private circularity platforms on plastic materials, waste separation and recycling
		Plastic waste collection & handling	• Operational cost reduction through separate waste collection • Increased availability of quality data on plastic waste collection • Increased circularity platforms on separate waste collection
		Plastic waste sorting	• Operational cost reduction through sorting plastic waste • Product differentiation due to processing sorted waste • Increased circularity platforms for sorted plastic waste

(Continued)

TABLE 9.1 (Continued)

Metric to advance circularity	Metric objectives	Applicable value chain stages	Performance indicators per stage
		Plastic recycling	• Operational cost reduction through processing plastic recyclables • Product differentiation due to processing recyclables • Increased availability of quality plastic recycling data • Increased circularity platforms for sustainable plastic recycling
		Plastic recyclate market	• Operational cost reduction in supplying plastic recyclates to the market • Increased availability of quality data on recyclate supply and demand • More circularity platforms on market regulation and development
Policy and incentives to promote circularity	• Reduce research and development costs towards advancing the circularity of the system • Promote the economy (supply and demand) of reduced material, reuse model, and recycled plastics to advance plastic system circularity	Plastic production	• Increased use of plastic recyclate in production • Increased number of plastic re-manufacturing and recycling organisations and job opportunities • Increased re-manufactured and recycled plastic produced
		Plastic product manufacturing	• Increased number of product manufacturers and production rate • Increased number of plastic re-manufacturing and recycling job opportunities • Increased re-manufactured and recycled plastic products manufactured
		Plastic waste generation	• Increased consumers are separating their plastic waste • Increased plastic waste separated

TABLE 9.1 (Continued)

Metric to advance circularity	Metric objectives	Applicable value chain stages	Performance indicators per stage
		Plastic waste collection & handling	• Increased plastic waste collected (collection rate) • Increased plastic waste collection job opportunities
		Plastic waste sorting	• Increased plastic waste sorted • Increased plastic waste sorting job opportunities • Increased quality of plastic waste sorted
		Plastic recycling	• Increased plastic waste recyclers and recycling rate • Increased plastic recycling waste job opportunities • Increased quality recyclates
		Plastic recyclate market	• Increased dealers operating in the market • Increased job opportunities in the market • Increased quality recyclates brought to the market
Improved technical competence and expertise	• Develop skills and expertise in the chemistry and product design of plastics to continuously advance plastic circularity	Plastic production	• Increased recycled plastics and re-manufactured products • Increased high-quality recycled plastics produced • Increased funding of R&D expenditure for circular plastic material processing • Increased operators with certificates in recycled plastic chemistry and quality

(Continued)

TABLE 9.1 (Continued)

Metric to advance circularity	Metric objectives	Applicable value chain stages	Performance indicators per stage
	• Efficient and sustainable re-manufacturing, repairing, sorting, and recycling innovations, technologies, and operations towards increased circularity	Plastic product manufacturing	• Increased recycled plastic products manufactured • Increased high-quality recycled plastic products manufactured • Increased R&D expenditure for circular product design knowledge
		Plastic waste collection & handling	• Increased sustainable plastic waste collected • Time reduction in collection
		Plastic waste sorting	• Increased plastic waste sorted • Time reduction in sorting • Increased sorters with skills and training certificates in plastic (types) waste sorting
		Plastic recycling	• Increased recyclates • Increased high-quality recyclates • Increased funding of R&D for plastic waste recycling • Increased recyclers with skills and training certificates
		Recyclate market	• Increased quality recycled products in the market • Increased dealers with skills and training certificates in the market
Improvement in recovery value	• Increase circularity through the production of high-quality plastic recyclables with acceptable properties	Plastic recycling	• Increased qualitative recyclables • Constant and consistent satisfaction of the average quality coefficient or standard units of measurement for good recyclates • Competitive in average selling price of the recyclate to virgin plastic • Reduced scrap production of recyclates • Reduced consumer complaints (warranty claims or outright rejection) regarding poor recyclates

		Plastic recyclate market	• Competitive recyclate average selling price to virgin plastic in the market • Reduced consumer complaints (in warranty claims or outright rejection)
Secure demand	• Promote circularity for plastics by facilitating the use of used, repaired, repurposed and recycled plastics and plastic materials • Security of buyers of used, repaired, repurposed and recycled plastics	Plastic production	• Increased consumption of produced circular plastics • Growth in sales revenue of circularly produced plastics • Growth in the volume of plastic produced for products
		Plastic product manufacturing	• Increased consumption of circularly manufactured plastic products • Growth in sales revenue of produced plastic products • Growth in the volume of circular plastic products
		Plastic waste sorting	• Increased use of sorted plastic waste for recycling • Growth in sales revenue of sorted plastic waste
		Plastic recycling	• Increased users of recyclates for production or consumed recyclates • Growth in sales revenue of recyclates • Growth in the volume of produced recyclates
		Plastic recyclate market	• Increased buyers or purchased recyclates • Growth in sales revenue of recyclates • Growth in recyclates supplied to the market • Increased markets, selling points or distributors of recyclates
Awareness	• Increase in improved consumers' decisions and efforts to participate in and promote plastic reduction,	Plastic waste generation	• Increased plastic waste separated by consumers • Increased inquiries from consumers regarding plastic waste separation • Increased consumer participation level in surveys and information

(Continued)

TABLE 9.1 (Continued)

Metric to advance circularity	Metric objectives	Applicable value chain stages	Performance indicators per stage
	reuse, repair, repurposing and plastic waste recycling		
Essential infrastructure	• Improvement in facility and capacity to manage and increase circularity practices, processes, and activities within the plastic system	Plastic waste collection & handling	• Increased plastic waste collection facilities • Increased project investment and financing for collection facilities
		Plastic waste sorting	• Increased sorting facilities • Increased project investment and financing for sorting facilities
Better management of waste generation scale	• Improve and increase sorting and depositing waste to enable and drive the collection of plastic waste in huge quantities for increased recycling and circularity • Supply security of adequate amount of sorted plastic waste for increased recycling and circularity of plastics	Plastic waste generation	• Increased plastic waste generated by consumers
		Plastic recycling	• Increased plastic waste generated

Source: Modified from Olatayo, K. I., Mativenga, P. T. & Marnewick, A. L. (2023). Plastic value chain and performance metric framework for optimal recycling. Journal of Industrial Ecology, 27(2), 601–623. https://doi.org/10.1111/jiec.13384.

The different levels contain processes that need to be implemented by that system or organisation for growth. Also, they represent hierarchical progressions from the elementary or least level of relatively poor processes and practices to the highest level of best, well-defined and systematic processes and practices (Wendler, 2012). The improvement in processes and practices of a plastic system will result in promotions in levels up the maturity model, making it an enabler for realising circularity and optimal development of the system. While the developmental levels of maturity models are fundamental components of the maturity model, the range of criteria or elements defining the improvement process is equally important.

Maturity models have been categorised in terms of the various dimensions they relate to or address, including people maturity, process maturity and technology maturity (Mettler, 2011). People's maturity describes the different levels of attainment of knowledge development and proficiency improvement by individuals. Process maturity relates to different levels of clearly defining, measuring, managing and controlling a process. Technology maturity can be described as the different levels of threshold reached by a technology.

Different maturity models have been developed for evaluating systems or organisations in various fields or disciplines, emphasising the increasing popularity, acceptance, suitability and usefulness of the tool. The various fields where maturity models have been applied include supply chain management, project management, product development, energy management, software development and technology. Their application to the plastic system enables the development of the system and improved circularity of the material.

The Quality Maturity Grid was developed by Crosby in 1979, and it formed one of the roots of maturity models (Fraser et al., 2002). The model was developed to evaluate the maturity of organisations, and it assessed organisational processes based on five levels of maturity. These maturity levels, in order of ascension, are: stage 1 (uncertainty), stage 2 (awakening), stage 3 (enlightenment), stage 4 (wisdom) and stage 5 (certainty) (Fraser et al., 2002).

The Supply Chain Process Maturity Model was developed by de Oliveira et al. (2011), using a questionnaire survey and statistical analysis (exploratory factor analysis). The model was structured into five different levels using groups of supply management process capability indicators, which include foundation building and order management, production planning and scheduling, distribution network management, procurement team, demand management and forecasting process governance, strategic planning team, strategic behaviour, responsiveness and collaboratively integrated practices, supply network management and customer integration. Similarly, the five levels are level 1 (foundation), level 2 (structure), level 3 (vision), level 4 (integration) and level 5 (dynamic).

Lockamy and McCormack (2004) developed the Supply Chain Management Maturity Model to evaluate the relationship or link between supply chain management maturity processes and supply chain performance. Developed by combining the business process orientation maturity model concept, interviews with supply chain industry experts and supply chain survey data, the model comprises five different levels. These levels are defined by various characteristics associated with process maturity.

The Eco-design Maturity Model was developed by Pigosso et al. (2013). It was designed specifically for the promotion of eco-design processes and practices and environmental sustainability, with respect to the development of products and related processes. Structured into five levels (incomplete, ad hoc, formalised, controlled and improved) in ascending order of improvement, the focus of the model is to limit the impacts of the product on the environment throughout its life cycle, without compromising cost, quality, functionality and performance.

The development of the maturity model for circular plastic systems is an essential contribution with numerous benefits. These include evaluating an existing case or situation for the development of a control and improvement strategy, plan or roadmap; identifying benchmark systems of similar or comparable processes or practices in other environments or settings; equating the maturity level of processes of a system with that of best practices; and evaluating the adequacy of resources towards adopting required actions with a system.

As represented in Table 9.2, the maturity model for circularity of the plastic system has five levels of maturity, which are, in increasing order of improvement: level 1 (ad hoc and unstructured), level 2 (structured), level 3 (visionary), level 4 (integrated) and level 5 (connected and dynamic).

The approach for the development of this maturity model for circular plastic economy followed after the value chain coordination maturity model for plastic material recycling developed by Olatayo et al. (2023), wherein all the different maturity levels in the model are guided by seven maturity elements, including chain actors' integration, customer satisfaction, performance targets, process costs, quality data and database, quality recyclables and products, structured processes, and system performance review, for the circularity evaluation of the system. Chain actors' integration defines a situation where value chain actors are identified and defined and related organisations seek practices, processes, and activities that are collaborative and integrative among all the actors or stakeholders within the plastic value chain. Customer satisfaction describes a consumer-driven system and one that is responsive to the satisfaction of customers and process improvements. Performance targets ensure targets are set and realised consistently for process performance. Process costs describe substantial reductions in costs relating to processes, practices, and activities within the value chains. Quality data and databases demonstrate that there is availability of quality and credible data on the mass flow of plastic material, plastic products, and plastic waste across the value chain. Quality recyclables and products signify that high-quality plastic materials, plastic products, sorted plastic waste and plastic recyclables flow across the value chain. Structured processes relate to value chain practices, processes, and activities being defined, structured and documented formally.

TABLE 9.2 Plastic circularity maturity model for plastic systems.

Maturity level	Title	Characteristics
5	Connected and dynamic	• Integrative and collaborative practices between value chain actors towards circularity of the system • Systematic tracing of plastics, plastic waste and recyclables in the plastic system for improved circularity across the value chain • Responsiveness of the recycling system to changes in the recyclate market in advancing circularity of the system • Responsiveness of the re-manufacturing system to changes in the market in advancing circularity of the system • Value chain management within multi-organisations, with common objectives to achieve circular economy for plastics • Reliable and quality data and database on mass flow of plastic waste, plastic material and plastic products exist and are available towards advancing circularity across the value chain • Embedment of value chain in trust and mutual dependency for circularity of the plastic system • Performance and maturity measurement of the plastic system regarding improved plastic circularity • Collaborative investments to improve the circular plastic system and share returns • Sorted plastic waste, recyclables, materials and plastic products of high quality towards achieving improved circularity across the value chain • Responsive and consumer-focused circular plastics value chain
4	Integrated	• Organisations are in pursuit of a collaborative environment for all value chain actors towards advancing circularity across the value chain • Comprehensive supply and demand forecasts for plastic material, products, waste and recyclables start to take form for improved circularity of the system, as consumers' orders are treated distinctly • Towards advancing circularity of the system, analytics are beginning to be applied in collating quality and reliable data on the mass flow of plastic material, products and waste across the value chain • Circularity processes and practices across the value chain define organisational structures and jobs • Value chain traditional practices start to be discarded, as circularity practices take root • Set targets for system performance towards achieving circularity across the value chain that are consistently met

(Continued)

TABLE 9.2 (Continued)

Maturity level	Title	Characteristics
		• Significant reduction in the costs associated with circularity strategies, processes and practices across the value chain • Consumer satisfied and included in the drive for process improvements towards advancing circularity across the value chain
3	Visionary	• Formal set-up of platforms for the development and management of circularity strategies such as sustainable product design, reuse, repair, repurpose, recycling and re-manufacturing activities • Owners of value chain processes and practices that can advance the circularity of the system are established and tasked with each process's performance • Process performance targets towards circularity of the system are usually achieved • Start of cost reduction with respect to circularity practices and processes across the value chain • Consumer satisfaction starts to be noticed, as consumers are included in process improvements to implement circularity across the value chain
2	Structured	• Structuring of the value chain processes towards circularity of the system starts • Basic value chain processes towards circularity of the system are defined and documented • Organisation practices and jobs are still traditional, rather than circular • Quality data availability to assess and advance circularity of the system remains low across the chain • Process performance targets towards circularity of the system are defined, but are occasionally achieved • Traceability of plastics, plastic waste and recyclables to improve circularity of the chain remains low • Costs related to circularity processes and practices across the value chain are still high • Understanding of the principles and advancement of circularity of plastics is still low, but improving • Consumer satisfaction remains low, although improving due to a gradual improvement in service

1	Ad hoc and unstructured	• Planning to create a basic structure for circularity processes and practices across the value chain • Identification of value chain actors to promote circular model for the plastic system • Plastic circularity processes and practices regarding collection, sorting, operations, sales, logistics and other related activities are not properly defined, structured and documented • Different process metrics to assess the circularity of the value chain are missing • Low traceability of plastics, plastic waste and recyclables for improved circularity of the value chain • Process performance targets for system circularity are largely undefined, or are unrealised if they are defined • High cost of circularity practices and processes across the value chain • Low-quality plastic waste, recyclables and products due to poor circularity practices • Mostly unavailable quality data and database to assess and advance circularity across the value chain • Separate efforts in advancing plastic circularity rather than functional engagement and collaboration • Low understanding of the principles and advancement of circularity of plastics • Low consumer satisfaction within the value chain due to poor services

Source: Modified from Olatayo, K. I., Mativenga, P. T. & Marnewick, A. L. (2023). Plastic value chain and performance metric framework for optimal recycling. Journal of Industrial Ecology, 27(2), 601–623. https://doi.org/10.1111/jiec.13384.

System performance review describes the frequent auditing of the reliability and performance of the plastics value chain system.

The development of the maturity model for circularity of plastics was informed by the Supply Chain Process Maturity Model (de Oliveira et al., 2011) and the Supply Chain Management Maturity Model (Lockamy and McCormack, 2004). The conceptual model integrates components of the two maturity models and the processes and practices across the plastics value chain. The features of the two models and their maturity levels are largely comparable, aligned and relevant, despite the differences in the names of the levels. Also, these features are largely applicable to the processes and practices that take place in the plastic system. The analysis, synchronisation and extrapolation of each equivalent level of the two models and their level-by-level corresponding features, together with the plastics value chain processes, produced the respective maturity levels and features of the conceptual maturity model for the plastic system.

The maturity model for circularity of plastics presents essential processes, practices and activities that require adoption and implementation across the plastics value chain for the optimal development and maturity of the system towards a circular model. The maturity levels represent advancement of processes and practices towards a mature and optimally developed plastic system. These levels are centred on capability, predictability, control, efficiency and effectiveness, similar to the two maturity models analysed. Detailed descriptions of the maturity levels are presented in the following.

9.3.1 Ad hoc and unstructured

At this level, the plastics value chain is still undefined and unstructured. The various stakeholders operating across the different stages of the value chain are not yet collaborating. There is an individualistic operation (working in silos). These identified stakeholders start to make plans to put in place basic structures for the processes in the value chain, as these processes are yet to be properly defined. Due to the system being unstructured, there is no process measurement, there are cost-intensive processes, low traceability of plastic materials, waste and recyclables, poor quality of plastic recyclables and products, poor collection and collation of data across the value chain, and missed performance targets. Furthermore, poor services are rendered to consumers, households and communities, which leads to low customer satisfaction.

9.3.2 Structured

At level 2, the structuring, definition and documentation of the basic processes of the plastics value chain have started, although these processes and activities might still be traditional and not sophisticated. Considering that structuring is just beginning, the collection of quality data by the various stakeholders across the value chain still remains largely low, and the cost of operations across the system still remains high. However, the system has begun recording a more predictable process performance, as performance targets are now being realised, albeit intermittently. Customer satisfaction, while still low, has started improving.

9.3.3 Visionary

This level sees plans integrating various processes of the plastics value chain being formalised. Activities are becoming more structured and defined and development and management of the value chain are taking root, as the relevant stakeholders in the different value chain stages are established and taking ownership of their processes. These stakeholders are starting to engage, collaborate and share data and information for the development of the system. The formalisation of structure begins to result in frequent achievement of process performance targets, the beginning of a reduction in the costs of processes and activities of the value chain and the inclusion of consumers in process improvement to enable customer satisfaction.

9.3.4 Integrated

At this level, the plastic system has become an establishment for a more collaborative environment for all stakeholders operating across the value chain, as they cooperate in processes, planning and projections. Traditional processes and practices are beginning to reduce across the chain. The system starts to experience a comprehensive supply and demand forecast for plastic material products, plastic waste and recyclables. This enables the satisfaction of consumers, as their inclusion in process improvements is promoted. The application of analytics in the collation of quality data on plastics and the corresponding waste starts, while performance targets for the plastic system are now met consistently. Furthermore, considerable cost reduction in processes of the value chain is recorded.

9.3.5 Connected and dynamic

In level 5, as the plastic system is now well-defined and structured, stakeholders are not only collaborative across the value chain, but also integrated in goals, plans, strategies, processes and control. They also witness trust, openness and mutual dependency among themselves, while customer-focused collaboration and responsiveness are entrenched in the system. All participants in the value chain have shared values and investments in the plastic system improvement. As a result, multi-organisation value chain management takes place, high-quality plastic recyclables, materials and products are produced at relevant stages of the value chain, quality data on plastics and plastic waste is collated and shared for system planning and development, circularity practices are improved and the performance of the system and the respective stages is measured.

9.4 Chapter summary

This chapter focused on contributing to improved circularity for plastics and advancing the plastic system to world-class standards by developing a performance metric framework for assessing the performance of the system. This involved identifying key performance metrics across the plastics value chain. Additionally, a maturity model was developed for the assessment and referencing of the growth and development of plastic systems in relation to the key performance metrics.

The complete circularity and optimal performance of the plastic system require the pathways to development and the circular economy for plastics to be improved continuously. The assessment of the performance and development of the plastic system helps in the design of action plans to bridge the gaps between the current level of performance and maturity and the desired optimal system performance and full maturity.

The performance metric framework for the circular economy of plastics comprises a set of key metrics for assessing the level of performance of a plastic recycling system and benchmarking its success towards optimal circularity performance, regardless of geographical location. The key metrics on which the framework is centred represent critical factors required for increased plastic circularity and system performance. They include policy and incentives, value chain coordination, infrastructure, technical expertise, demand, waste generation scale, recovery value and awareness. This broader framework integrates the key performance metrics, objectives of the metrics for the plastic system, assessment methods for measuring the respective metrics, the applicable or relevant stages of the plastics value chain for each performance metric and the performance indicators to assess the relevant performance metrics. The application of the performance metric framework to any plastic system to advance the circular plastic economy involves assessing each stage of the value chain through the stage-related performance metrics and corresponding performance indicators. It helps determine the individual performances of these stages and identify specific stages that require improvement.

Furthermore, a maturity model for the circular plastic economy was developed to evaluate the level of development or maturity of any plastic recycling system. The proposed circularity model has five levels of maturity, which are, in increasing order of improvement, level 1 (ad hoc and unstructured), level 2 (structured), level 3 (visionary), level 4 (integrated) and level 5 (connected and dynamic). The maturity levels in the model are guided by seven maturity elements for the assessment of the system: chain actors' integration, customer satisfaction, performance targets, process costs, quality data and database, quality recyclables and products, structured processes and system performance review. The different maturity levels represent advancement of processes and practices towards a mature and optimally developed circular plastic system.

References

Antunes, P., Carreira, P., Mira da Silva, M., 2014. Towards an energy management maturity model. Energy Policy 73, 803–814. Available at https://doi.org/10.1016/j.enpol.2014.06.011.

Fraser, P., Moultrie, J., Gregory, M., 2002. The use of maturity models/grids as a tool in assessing product development capability. IEEE International Engineering Management Conference, Cambridge, UK, vol.1. IEEE, pp. 244–249. Available at https://doi.org/10.1109/IEMC.2002.1038431.

Lockamy, A., McCormack, K., 2004. The development of a supply chain management process maturity model using the concepts of business process orientation. Supply Chain Management: An International Journal 9 (4), 272–278. Available at https://doi.org/10.1108/13598540410550019.

Mettler, T., 2011. Maturity assessment models: a design science research approach. International Journal of Society Systems Science 3 (1/2), 81. Available at https://doi.org/10.1504/IJSSS.2011.038934.

Olatayo, K.I., Mativenga, P.T., Marnewick, A.L., 2023. Plastic value chain and performance metric framework for optimal recycling. Journal of Industrial Ecology 27 (2), 601–623. Available at https://doi.org/10.1111/jiec.13384.

de Oliveira, M.P.V., Bronzo, M., Kevin, P., 2011. The Supply Chain Process Management Maturity Model – SCPM3. In: Onkal, D. (Ed.), Supply Chain Management - Pathways for Research and Practice. InTech, pp. 201–218. Available at https://doi.org/10.5772/18961.

Pigosso, D.C.A., Rozenfeld, H., McAloone, T.C., 2013. Ecodesign maturity model: A management framework to support ecodesign implementation into manufacturing companies. Journal of Cleaner Production 59, 160–173. Available at https://doi.org/10.1016/j.jclepro.2013.06.040.

Pöppelbuß, J., Röglinger, M., 2011. What makes a useful maturity model? A framework of general design principles for maturity models and its demonstration in business process management. Proceedings of the 19th European Conference on Information Systems (ECIS 2011) (Paper 28). Association for Information Systems, pp. 1–28.

Wendler, R., 2012. The maturity of maturity model research: A systematic mapping study. Information and Software Technology 54 (12), 1317–1339. Available at https://doi.org/10.1016/j.infsof.2012.07.007.

Chapter 10

Case studies on country transition to a circular plastic economy

10.1 The global plastics value chains and transition towards circularity

The global plastics value chain incorporates the complete life cycle of plastics, ranging from raw material extraction for plastics production to the eventual disposal of the corresponding waste after consumption. It equally involves various stakeholders across the value chain. The plastics value chain globally is complex. While plastics are valuable and widely used, and the value chain generates revenue and provides jobs, but countries are facing challenges in managing this value chain, specifically regarding the issues of plastic waste and pollution.

Responding to these plastic waste and pollution problems requires a shift towards the circular economy for plastics. The circular plastic economy is gaining momentum, as more countries and institutions are initiating and implementing circularity strategies across the value chain, targeting sustainable product design, reuse, repair, repurpose, remanufacturing and recycling of plastic waste. The shift to plastic circularity will require circularising all the respective stages of the value chain, rather than just addressing individual stages separately.

In achieving the transition to circularity of the plastic system, circularity targets are required to stimulate plastics value chains towards a defined and shared goal. Circularity targets are set in relation to achieving a certain goal and maturity on the waste hierarchy. This can be, for example, a time-bound plan for reaching a zero landfill, or defined plastic collection, recycling or remanufacturing rates. Pathways are then required to transition towards the defined target.

For practical insights and understanding of transitioning towards circularity of the plastics value chain, case studies were examined. This involves reviewing the circularity targets and measures adopted by international organisations and some countries, in relation to the flow of plastics and plastic waste. The review includes the different policy interventions of the government of these case study countries and their progress towards advancing the circular plastic economy. Table 10.1 presents a comparative assessment of the different circularity targets

Plastic Circularity. DOI: https://doi.org/10.1016/B978-0-443-40667-6.00012-3

TABLE 10.1 Case studies on the transition towards a circular economy for plastics.

	Circularity target initiative	Target year	Target scope	References
United Nations	UNEP Plastics Initiative	2030	• Target 1 – 45 Countries will have improved their plastics policies and regulations to enable robust transition towards a just and safe circular economy of plastics. • Target 2 – 500 Private sector actors in key industry sectors will have adopted effective circular solutions and innovations, committed to circular economy to reduce plastic pollution. • Target 3 – 50 Financial Institutions will have included circularity or plastic pollution prevention and reduction policies in their sustainability strategy and/or targets.	United Nations Environment Programme (2025)
European Union	European Plastics Pact Roadmap	2025	• Target 1 – Design for reusability and recyclability: Design all plastic packaging and single-use plastic products placed on the market to be reusable where possible and, in any case, recyclable by 2025. • Target 2 – Responsible use of plastics: Move towards a more responsible use of plastic packaging and single-use plastic products, aiming to reduce virgin plastic products and packaging by at least 20% (by weight) by 2025, with half of this reduction coming from an absolute reduction in plastics.	Waste and Resources Action Programme (2020)

			• Target 3 – Collection, sorting and recycling: Increase the collection, sorting and recycling capacity by at least 25 percentage points by 2025 and reach a level that corresponds to market demand for recycled plastics. • Target 4 – Use of recycled plastics: Increase the use of recycled plastics in new products and packaging by 2025, with plastics user companies achieving an average of at least 30% recycled plastics by weight in their product and packaging range.	
South Africa	National Waste Management Strategy	2050	• Prevent waste, and where waste cannot be prevented, ensure 40% of waste is diverted from landfill within 5 years; 55% within 10 years; and at least 70% within 15 years, leading to Zero-Waste going to landfill.	Department of Environment, Forestry and Fisheries (2020)
	South African Plastics Pact	2025	• Target 1 – Taking action on problematic plastic packaging. • Target 2 – 100% of plastic packaging to be reusable, recyclable or compostable. • Target 3 – A minimum of 70% of plastic packaging effectively recycled. • Target 4 – A minimum of 30% recycled content across all plastic packaging.	South African plastics pact (2021)

(Continued)

TABLE 10.1 (Continued)

	Circularity target initiative	Target year	Target scope	References
Nigeria	National Action Roadmap Towards a System Change in Plastic Pollution: A Roadmap for the Sustainable Use of Plastics in Nigeria	2040	• Up to 58% increase in circularity by 2040 with decisive interventions, up to 67% decrease in plastic pollution by 2040 following a system-change scenario. • Plastic waste recycling increased from 10% to 31% • Safe disposal methods rise to 12%. • Mismanaged plastic waste is expected to decrease to 30% of the total plastic waste generated.	GPAP (2024)
Egypt	Waste 50 by 2050	2050	• Recycle 50% of waste generated in Africa.	Egypt Today Staff (2025)
India	India Plastics Pact	2030	• Define unnecessary or problematic plastic packaging and address by redesign and innovation. • 100% of plastic packaging to be reusable, recyclable or compostable. • Effective recycling of plastic packaging should be 50% • Average recycled content for all plastic packaging should be 25%.	Confederation of Indian Industry, & World Wide Fund for Nature–India (2022)

China	5-Year Action Plan (2021–25):	2025	• Phasing out Single-Use Plastics: China is phasing out single-use plastics in major cities, including straws, bags, and thin plastic film used in agriculture. • Circular Economy Focus: China is working towards a circular economy for plastics, focusing on eliminating unnecessary packaging, increasing reusable packaging, ensuring recyclability, and increasing the use of recycled content. • International Cooperation: China is actively participating in international efforts to combat plastic pollution and is engaging in discussions for a global treaty on plastic pollution.	Yee (2023)
Germany	National Circular Economy Strategy	2045	• Achieve the EU target of doubling the share of secondary raw materials in the total amount of raw materials used by 2030. • Achieve at least 25% of the demand for strategic raw materials through recycling. • Reduce per capita volume of municipal waste by 10% by 2030 and by 20% by 2040. • Halve per capita raw material consumption by 2045.	Weiner et al. (2025)

(*Continued*)

TABLE 10.1 (Continued)

	Circularity target initiative	Target year	Target scope	References
United Kingdom	UK Plastics Pact	2025	• 100% reusable, recyclable, or compostable plastic packaging. • 70% of plastic packaging effectively recycled or composted. • Eliminate single-use packaging. • 30% average recycled content across all plastic packaging.	Wrap (2024)

and proposed measures by the United Nations (UN), European Union (EU), and seven countries, including South Africa, Nigeria, Egypt, India, China, Germany, and the United Kingdom (UK).

As expressed in the table, the UN, through its affiliate United Nations Environment Programme, initiated the Plastics Initiative for governments and the private sector to transition towards a circular system for plastic. Targets set for these goals by 2030 include the improvement of plastics policies in 45 countries towards transitioning to plastic circularity; adoption of innovative and circular solutions to reduce plastic pollution and increase circularity by 500 actors in the private sector; and the inclusion of policies promoting plastic circularity and pollution prevention and reduction in the sustainability targets and/or strategy of 50 financial Institutions. Proposed measures to the goals' achievement include support for businesses to develop circular business models and sustainable product design, development and implementation of broad policies across the plastics life cycle, and behavioural change, among other measures.

The EU introduced the Land Directive targets to promote the progressive reduction of recyclable or recoverable waste ending up in landfills in a bid to increase plastic circularity. It specifically mandated zero landfilling of all recyclable or recoverable waste by 2030, except where landfilling offers the most environmentally sustainable outcome. Also, it is expected that the share of municipal waste landfilled will be reduced to 10% by 2035. With the main goal of advancing the circular economy through improving waste management, increasing recycling and reducing landfilling, it provides measures towards realising the objective, particularly the use of economic instruments and other incentive measures to promote the application of the waste hierarchy options, including prevention; preparing for reuse; recycling; other recovery; and disposal, in order of priority. In addition, the European Plastics Pact Roadmap was introduced to achieve, by 2025 the design for reusability and recyclability for all plastic packaging and single-use plastic products; responsible use of plastic packaging and single-use plastic products; increase the collection, sorting and recycling capacity; and increase the use of recycled plastics.

A comparative analysis of the different circularity target initiatives of the case study countries shows that only India had the same target year as the UNEP. The European Plastics Pact Roadmap, South African Plastics Pact, UK Plastics Pact, and the 5-Year Action Plan had the same target year of 2025, which is below the UNEP. The target years for the National Waste Management Strategy, National Action Roadmap Towards a System Change in Plastic Pollution, Waste 50 by 2050, and the National Circular Economy Strategy of South Africa, Nigeria, Egypt, and Germany, respectively, are beyond that of the UNEP. However, the target scopes of the different initiatives broadly align with those of the UNEP, and are also largely in alignment with each other. Generally, they propose critical circularity measures such as sustainable product design, prevention and reduction of waste, reuse, repair, separation of waste and an increase in recycling. Furthermore, as proposed by the UNEP Plastics initiative, the roles and responsibilities of stakeholders were absent in the plastics pact. Previous chapters of this book have emphasised how critical stakeholders are to achieving set targets towards the transition to plastic circularity. Also, the modelling of pathways to plastic circularity provides opportunities for stakeholders to make informed and knowledge-supported choices in achieving circularity. These submissions totally align with the recommendation of the UNEP for the involvement of the government, private sectors, and financial institutions in achieving a circular economy for plastics.

10.2 South Africa

10.2.1 Plastics production and products manufacturing

South Africa produces plastic monomers and polymers in substantial quantities, with SASOL as the foremost producer, and other producers such as Safrisol supplying to the local and international polymer markets (Mofo, 2020). The industry for polymer production is well established and has a structure, with access to raw materials for plastics production. The plastic products manufacturing sector is also well established and has sizable manufacturing capacities and capabilities, although high operational costs, negligible advanced manufacturing, inadequate skilled human resources and slow upgrade of technology represent challenges to the sector (Mofo, 2020). There is some level of engagement and collaboration between the sectors of polymer production and product manufacturing, as the polymer sector feeds the local market for plastic product manufacturing (Plastics SA, 2022). The annual primary plastic and plastic products per capita production were 25 and 32 kg/capita, respectively, in 2021 (Olatayo et al., 2024b).

Overall, both the primary plastic production and product manufacturing sectors were rated largely as "visionary" in terms of maturity (Olatayo et al., 2024a). Maturity describes the level of development or growth of the sectors in the plastics value chain system. "Visionary" (level 3, out of 5 in ascending order of development) represents the "*existence of established owners of various processes of the plastics value chain, who are tasked for performance; targets for performance are mostly realised; process costs start to reduce along the value chain; and consumers' inclusion in the improvement of processes starts to produce customer satisfaction*".

10.2.2 Plastics use and waste generation

The annual primary plastic and plastic products per capita consumption were 32 and 35 kg/capita, respectively, in 2021, and the post-consumer plastic waste generated (excluding import, export and production waste) in the same year was 26 kg/capita (Olatayo et al., 2024b). This stage is characterised by illegal open dumping, littering, minimal source separation of waste and a lack of structure and incentives to separate waste at source (Greencape, 2019). National campaigns and programmes to create awareness on separation at source are equally non-existent, although source separation occurs in some metropolitan areas (Department of Environment, Forestry and Fisheries, 2020). As expected, the low separation at source and the absence of structures for this result in poor quality of the waste for processing and high operational costs. The sector was rated "structured" (level 2) in terms of maturity (Olatayo et al., 2024a), which indicates that the *"processes of the value chain begin to be structured; elementary processes are being defined and documented; traditional jobs and practices still exist; quality data pertaining to the system is available in little quantity; targets for performance are rarely realised, even though defined; and high process costs still exist"*.

10.2.3 Plastics waste management and recycling

The South African national plastic waste flow analysis in Fig. 10.1 shows that of the total 1,810,194 tonnes of plastic waste generated in 2021, about 30% was recovered for recycling, 39% was landfilled, and 31% was littered or undisposed (Olatayo et al., 2024b).

With respect to shifting towards circularity for plastics, the system shows that more still needs to be done in the area of waste service delivery, separate waste collection, an integrated waste management system, investment in relevant infrastructure, and the enormous reliance on the informal sector for waste collection (Department of Environment, Forestry and Fisheries, 2020; Greencape, 2019). Also, engagement and collaboration among stakeholders within the sector need improvement (Jaarsveldt, 2016).

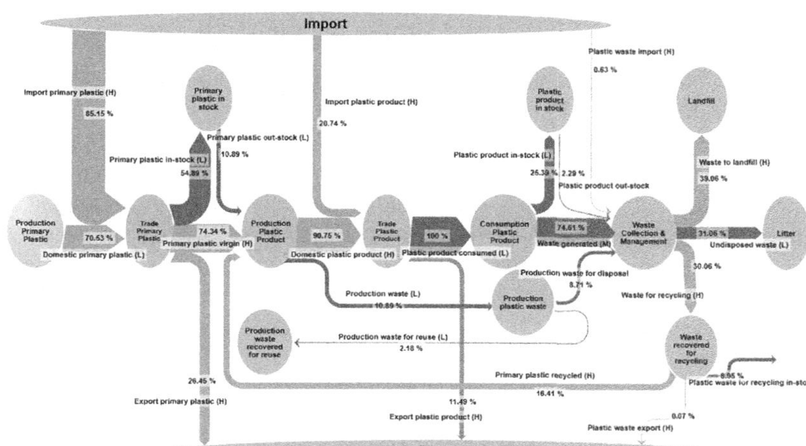

FIGURE 10.1 National plastics and plastic waste flow across South Africa in 2021. SanKey showing plastic and waste flow in South Africa.

The informal sector is a major player in the sector, as it makes up for the inadequacies in the collection and sorting of recyclables at source, helps divert waste away from landfills and supplies most recyclables for recycling (Department of Environment, Forestry and Fisheries, 2020). It recovers about 80%–90% of packaging waste for reuse and recycling (Plastics SA, 2017). The informal sector separates waste outside the source and recovers plastic recyclables from waste bins and landfills, selling them to buy-back centres, which then add value to the recyclables by sorting, cleaning and baling them, and subsequently selling to big recycling firms (Viljoen et al., 2019). While the recyclables through the buy-back centres are mainly well sorted, the processes are not technology-driven and are labour-intensive. Like many developing economies, the informal sector remains largely unregulated and marginalised despite the National Waste Management Strategy requiring the integration of informal waste collectors into waste management operations by municipalities.

Recyclers are similarly operating in an increasingly difficult environment, encumbered by rising process and electricity costs, as well as a general economic downturn (Jaarsveldt, 2016). Despite these deficiencies, industry-funded associations provide considerable support (Hoffman and Schencks, 2020). Investments in quality assurance laboratories and other equipment are enabling improved quality recyclates, notwithstanding the low recyclable separation at source. Regarding the maturity of this phase, the waste collection and sorting sector was rated "structured" (level 2) and the recycling sector was rated "visionary" (level 3) (Olatayo et al., 2024a).

10.3 Nigeria

10.3.1 Plastics production and products manufacturing

Nigeria was described as the sole producer of plastic polymer in the entire West African region, with about 498,000 tonnes produced in 2019 (World Bank, 2022). However, the plastics in the country are still largely met by imports, mostly from China. The growth of the sector is challenged by several factors, the foremost including limited availability of advanced technology and equipment, irregular and unreliable electricity supply for production, inadequate research and development, and poor funding and investment for large-scale production of plastic polymers and products (PwC, 2019).

10.3.2 Plastics use and waste generation

The consumption of plastics in the country is predominantly by the populace, but most of the subregions within the Economic Community of West African States (ECOWAS) depend heavily on the country for their plastics use. As of

2020 plastic packaging was the most consumed plastic product, representing a 54% share, followed by the construction sector with about 16% used in the industry (africon, 2024). The country was estimated to have a per capita consumption growth rate of approximately 3.6%.

10.3.3 Plastics waste management and recycling

This sector is characterised by non-separation or sorting of waste at the source of generation, which normally results in a collection of mixed waste streams. However, some plastic waste, such as polyethylene terephthalate (PET) bottles, is collected at waste bins, open dumps and landfills, and manually sorted by informal waste pickers for their reusability or repurposing (Nzeadibe and Ajaero, 2011). Some informal operators also purchase clean recyclables from households for reuse, repair, or repurposing.

The waste management system is generally characterised by a poor waste collection rate and a lack of advanced sorting. Uncollected waste across cities reportedly causes the blockage of water and drainage channels, leading to flooding (Duru et al., 2019). Inner towns and suburbs are usually inaccessible to waste collection trucks as a result of bad road surfaces and networks, limiting the collection of waste to easily accessible sections of towns and cities across the country. This scenario largely enables the indiscriminate and illegal disposal or dumping of waste in these inaccessible places, open spaces and water bodies. As a result, a significant share of the plastic waste generated ends up in open dumpsites and water bodies.

Recycling in Nigeria is mainly informal and not standardised. In the absence of a formal and well-regularised structure, informal waste recycling initiatives, though relatively well organised, facilitate the collection, separation, sorting, reuse and recycling of plastic waste. Mechanical recycling of plastic is practised in the country to a greater extent, whereas incineration is not widely accepted for the treatment of waste, except in limited cases such as medical waste (Ezeudu et al., 2024). Likewise, waste-to-energy solutions are rarely in practice.

10.4 Egypt

10.4.1 Plastics production and products manufacturing

Regarded as one of the most active and flourishing subsectors of the chemical industry in Egypt, the plastic system relies on domestic production and the import of plastic polymers. Domestic production was estimated at 2 million tonnes of different plastic types annually, contributing to about 24% of plastic raw materials needed for the manufacture of plastic products (Chemonics Egypt, 2021). The annual plastic import meets the remaining demand, at about 6 million tonnes.

10.4.2 Plastics use and waste generation

The total plastic consumption in Egypt is approximately 8 million tonnes annually (Chemonics Egypt, 2021). Egypt was rated the largest consumer of plastic polymer in Africa at 2.1 million tonnes in 2017 (Obrecht et al., 2019), consuming more than the total mass of consumption of other North African countries, including Morocco, Algeria, Tunisia and Libya. The country consumes 11% of the share of plastic products consumed in the Middle East (Chemonics Egypt, 2021).

10.4.3 Plastics waste management and recycling

The annual plastic waste generation in Egypt was about 3.6 million tonnes in 2019 (Chemonics Egypt, 2021), at 34 kg/cap, which requires a high-performing waste management system. Regrettably, the plastics waste sector is limited in effective waste collection services; this includes a lack of waste separation at source for the collection of quality plastic recyclables, complicated waste collection processes and practices, improper handling and transportation of waste and deficient transfer stations. However, the collection of waste can be fairly regular, and there is a secondary market for collected clean plastic recyclables.

The informal sector in Egypt, known as "Zabbaleen", is very vibrant and dominates the waste collection process, amidst the complicated waste collection system. With more than 50 years of experience, it plays a very active role in waste management and contributes significantly to the reuse and recycling of plastic waste (Chemonics Egypt, 2021). The efficiency and effectiveness of the informal collection sector surpass the efforts of the official system, thus emphasising its significance and supporting the call for this sector to be integrated into the official system. The sorting efficiency of the informal sector can be questionable, though, thus affecting the quality of recycled plastics.

Furthermore, the recycling system is challenged by mostly dilapidated equipment and inefficient processing techniques, which consequently lowers the quality and prices of recyclates and products (Chemonics Egypt, 2021). The sustained supply of plastic recyclables to formal recyclers is lacking, as the informal collectors control supply and can dictate where to supply or who gets what. The relationship between recyclers and users of the recyclates can equally be weak as a result of non-uniformity in the processed plastic recyclates. Limited expertise and know-how of product testing with respect to product standards and specifications is further regarded as fundamental for the growth of the plastic recycling system.

10.5 India

10.5.1 Plastics production and products manufacturing

India produced a total of about 19.3 million tonnes of plastics in 2018/19, with polyethylene (PE) accounting for 22% of the production (Emami et al., 2024).

There were over 30,000 companies manufacturing various plastics and plastic products in India in 2022, with 85%–90% of them being small and medium-sized enterprises (Nikam et al., 2022). Others are large manufacturing enterprises, but they are highly fragmented and independent of each other.

10.5.2 Plastics use and waste generation

The use of plastics in India rose from 14 to 20 million tonnes between 2016/17 and 2019/20 (Bardhan et al., 2024). The very high consumption, from a very large population, puts the country among the top five generators of plastic waste globally. Estimates indicate that about 10,829,000 tonnes of plastic waste were generated in 2023, equating to 7.6 kg/cap (EarthAction, 2024). Approximately 7,430,000 tonnes (70%) of the generated waste was mismanaged, which equated to 5.2 kg/cap.

10.5.3 Plastics waste management and recycling

In India, a substantial share of the plastic waste generated is poorly managed. The waste collection system is a major barrier, particularly in rural areas where waste is not separated or sorted, and the costs of waste collection and transportation are increasing (Nikam et al., 2022). A considerable portion of plastic waste is disposed of in open dumps, unregulated landfills and aquatic environments, causing damage to the ecosystem.

The plastic waste management system comprises both the formal and informal sectors. While the formal sector has to deal with several inadequacies in the management and recycling of plastic waste, the informal sector is rising to the occasion with its significant contribution to the system in the areas of waste collection, separation and plastic recyclable recovery and supply for recycling. The informal sector reportedly contributed to 70% of recycled PET bottles in the country (Shanker et al., 2023). The sector is largely praised for the putative high post-consumer plastic waste recycling rate of 60% (Bardhan et al., 2024). The formal recycling facilities are concentrated mainly in selected regions across India, but their capacity utilisation is below optimal performance, mostly due to the limited supply of quality recyclables, lack of advanced technology and lack of standardised processes and protocols (Singh, 2020).

10.6 China

10.6.1 Plastics production and products manufacturing

China is unarguably the largest producer of plastics globally. Over the years, its domestic plastic product output has continued to increase. This dropped in 2018 but rebounded in 2019, peaking at 81.84 million tonnes (Liu and Liu, 2023). A dynamic flow analysis shows that between 1980 and 2020, a total of

1.41 billion tonnes of primary plastics were produced (Liang et al., 2023). Some 70% of the country's plastic production is made up of the five most widely used polymers, with polypropylene (PP) and PE accounting for 60% of this (International Pollutants Elimination Network, & Shenzhen Zero Waste Environmental Public Welfare Undertakings Development Center, 2022).

10.6.2 Plastics use and waste generation

The country is not only the largest producer of plastics, but it also consumes the largest quantity of the products globally. An estimated 76.99 million tonnes of plastic products were consumed in 2019 (Liu and Liu, 2023), at 54 kg/cap. This resulted in a corresponding waste generation of about 63 million tonnes, at 44 kg/cap; 32%, 31%, 30% and 7% of this waste was landfilled, incinerated, recycled and leaked into the environment, respectively.

10.6.3 Plastics waste management and recycling

The level of development of the waste management system differs between the urban, developed rural and underdeveloped rural regions. In several underdeveloped rural regions, waste is randomly disposed of without treatment, unlike in the developed rural regions, where the system is gradually being integrated into the urban system (Liu and Liu, 2023). Waste collection also differs across different regions and areas in China. Collection in major cities for the upper class is quite adequate, whereas the poorer areas or communities experience basic services, particularly where waste is mixed (Zhang et al., 2010). Despite the differences in the collection rate for regions, collection is generally increasing across the country. The volume of urban and rural waste collection increased by 45% and 69%, respectively, between 2009 and 2018 (Liu and Liu, 2023).

In enabling the recycling of plastic waste, the separation of waste at source by households is relatively low, particularly in large cities where waste is mixed (Xiao et al., 2018). The informal sector is contributing to the separation of waste for recycling, though for economic purposes. The recycling of household plastic waste was categorised into two main systems: the renewable resources recovery system and the municipal solid waste (MSW) collection and transport system.

The renewable resources recovery system is the major path to plastic waste recycling, involving the informal waste collectors and recyclers. The MSW system uses mostly the sanitary landfill and incineration methods for waste treatment. A significant share of high-value recyclables is mostly collected by the informal sector and supplied to recyclers through the resource recovery system before it gets into the formal waste collection system, leaving the largely low-value recyclables in waste bins for the MSW collection and transport system (International Pollutants Elimination

Network, & Shenzhen Zero Waste Environmental Public Welfare Undertakings Development Center, 2022). The renewable resources recovery system is gradually evolving into large-scale businesses due to the integration of the sector. The recycling sector in China is not without its challenges, which include challenges in the regulation of the local recovery system, disproportionate tax burden, non-compliance by small recycling companies and overall unrefined industry management (International Pollutants Elimination Network, & Shenzhen Zero Waste Environmental Public Welfare Undertakings Development Center, 2022).

10.7 Germany

10.7.1 Plastics production and products manufacturing

Germany was the leading producer of plastics in Europe in 2023, accounting for more than 20% of the region's plastics demand. Approximately 13.8 million and 12.7 million tonnes were produced in 2022 and 2023, respectively (Statista Research Department, 2025). The drop in plastic production volume bet (European Environmental Agency, 2022) and 2023 impacted the production volume of Europe within the same years. should be The drop in plastic production volume between 2022 and 2023 impacted the production volume of Europe within the same years. Despite the downward trend in production between 2022 and 2023.

10.7.2 Plastics use and waste generation

Similar to production, the consumption of plastics declined between 2021 and 2023. The volumes consumed in 2021 and 2023 were 12.36 million tonnes and 11.25 million tonnes, respectively (Statista Research Department, 2025). The consumption quantity by the country in 2023 represents 22% of Europe's plastics demand. With the highest percentage in Europe, this emphasises the significance of the country in the region's plastics markets. 1.89 million tonnes of polyethylene made polymer the most commonly consumed plastic in 2023 (Statista Research Department, 2025). Other commonly used plastics in Germany include PP, accounting for 15.7% of the total plastics consumed in the same year. Flow analysis showed that approximately 6.28 million tonnes of plastic waste (at 76 kg/cap) were generated in 2019, with post-consumer plastic waste accounting for 5.35 million tonnes (at 64 kg/cap) of the total plastic waste (Lindner et al., 2019).

10.7.3 Plastics waste management and recycling

The waste collection service in Germany is functional. A significant percentage of households in the country experience good waste collection service, as separate waste collection is practised extensively (European Environmental Agency, 2022). The kerbside collection scheme is often used for the collection of waste, which promotes separate waste collection. Different-sized plastic bins and bags are provided at residents' doors and public places. The country generally has a culture of separating waste at source, which results in increased recyclable recovery (Nassour et al., 2018). Most households are also deposit refund scheme (DRS) participants. The country achieved a 74.8% ± 2.9% separate collection efficiency in 2017 (Picuno et al., 2021). The recycling sector is relatively established, becoming an integral part of the sector. An estimated 2.9 million tonnes of plastic waste generated in 2019 were mechanically recycled.

10.8 United Kingdom

10.8.1 Plastics production and products manufacturing

An estimated 1.7 million tonnes of plastics are produced annually in the UK (British Plastics Federation, 2025), putting the country among the top plastic producers and processors in the EU. Though highly dependent on polymer imports, the sector has continued to grow over the years. Recently, about half of the polymer quantity the country consumes annually is being produced domestically. An estimated 1.67 million tonnes of plastic polymer were produced in 2020 compared to the 3.3 million tonnes consumed the same year (British Plastics Federation, 2025).

10.8.2 Plastics use and waste generation

The UK is a relatively high consumer of plastics. In 2017, the country consumed about 6.3 million tonnes of plastics, with imports constituting a larger percentage of the consumption (Drewniok et al., 2023). Packaging and consumer goods constitute 2.2 million tonnes and 2.0 million tonnes, respectively. Estimates put the plastic waste generation for 2018 at about 5.2 million tonnes (at 78 kg/cap), with a projection of 6.3 million tonnes by 2030 (Elliot and Elliot, 2018).

10.8.3 Plastics waste management and recycling

Waste management in the UK is functional. Provision is made for separate waste collection of recyclables and residual waste by most local authorities. There is a regular alternate weekly collection of waste at kerbside for households and the bring system is applied (Hahladakis et al., 2018). Similarly, there is high separation of waste at source by households, with several residents utilising intervention schemes initiated by local authorities to meet their local recycling needs and the 50% recycling national target (Oluwadipe et al., 2022). Less than a third of generated post-consumer plastic waste is recycled, with more than half being burnt for the generation of energy. In 2022 20% of the UK's post-consumer plastic waste ended up at landfills (Alves, 2025).

10.9 Chapter summary

In this chapter, the global plastics value chain was reviewed with respect to the transition towards a circular economy for plastics. To review the value chain, the circularity targets and proposed measures by governments to advance circularity across the seven plastics value chain systems of different countries were considered, alongside those of the United Nations and the European Union. These countries are South Africa, Nigeria, Egypt, India, China, Germany and the United Kingdom. Furthermore, a general overview of the complex plastics value chain systems of the case study countries was discussed, which includes intricacies across the different value chain stages.

For South Africa, the production of plastics depends largely on coal and the sector is well established and structured, although high operational costs, negligible advanced manufacturing, inadequate skilled human resources and slow upgrade of technology are challenges. The consumption and generation of plastic waste are relatively high. The plastic waste management and recycling system is burdened with a poor waste collection and disposal system, poor quality of collected waste, enormous reliance on the informal sector for waste collection, and the price and perception of recyclates. Recyclers face rising process and electricity costs, as well as a general economic downturn, although investments in quality assurance laboratories and other equipment are enabling improved quality recyclates, notwithstanding the low recyclable separation at source. In shifting towards circularity, a number of policy mechanisms to advance circularity across the value chain have been initiated. However, despite the introduction of these circularity measures, compliance with and enforcement of these policies remain huge challenges.

Nigeria is the sole producer of resin in the entire West African region, but the plastics demand of the country is still largely met by imports, mostly from China. The sector faces challenges of limited availability of advanced technology and equipment, irregular and unreliable electricity supply for production, inadequate research and development, and poor funding and investment for large-scale production of plastic polymers and products. Plastic waste management and recycling are characterised by non-separation or sorting of waste at the source of generation, which normally results in the collection of mixed waste streams. The recycling sector is mostly informal and not standardised. The transition towards circularity of the value chain is not an area of focus. No known Act or regulation has been enacted to specifically address or advance the circular economy for plastics.

Egypt has one of the most active and flourishing subsectors of the chemical industry in the country, relying on the domestic production and import of plastic polymers. The country's consumption is more than the total mass of other North African countries. The annual plastic waste generation in Egypt is relatively high, with the plastic waste sector limited in effective waste collection services, which include a lack of waste separation at source for the collection of quality plastic recyclables, complicated waste collection processes and practices, improper handling and transportation of waste and deficient transfer stations. The informal sector dominates the waste collection process, in the midst of a complicated waste collection system and the recycling system, which is challenged by mostly dilapidated equipment and inefficient processing techniques. The transition towards circularity of the value chain shows that certain policies have been introduced to develop the waste management system, but more policy interventions and efforts at advancing the circularity of plastics across every stage of the value chain are required.

In India, there are over 30 000 companies manufacturing various plastics and plastic products, mostly involving small and medium-sized enterprises. The few large manufacturing enterprises are highly fragmented and operate in silos. The high plastic consumption puts the country among the top five generators of plastic waste globally. Unfortunately, a substantial share of the plastic waste generated is poorly managed. The waste collection system is a major barrier, particularly in rural areas where waste is not separated or sorted, and the costs of waste collection and transportation are increasing. A substantial share of plastic waste is disposed of in open dumps, unregulated landfills and aquatic environments, causing damage to the ecosystem. The formal recycling facilities are concentrated in selected regions across India, and their capacity utilisation is below optimal performance. The transition towards circularity of the value chain indicates that actions are being taken by union territories and states in the country to address plastic waste and promote circularity by implementing some form of circularity solution. At the national level, many policies have been initiated, but plastic regulation at that level remains relatively new and limited in scope and application. As with many developing economies, compliance with and enforcement of these national policies for desired results are lacking at both the local and regional levels.

China is unarguably the largest producer and consumer of plastics globally, which results in corresponding waste generation. The level of development of the waste management system differs between the urban, developed rural and underdeveloped rural regions. Similarly, waste collection differs across different regions and areas in China. Collection in major cities for the upper class is quite adequate, whereas the poorer areas or communities experience basic services. However, the waste collection rate is generally increasing across the country. The recycling of household plastic waste is categorised into two systems: the renewable resources recovery system and the MSW collection and transport system. The renewable resources recovery system is the major path to plastic waste recycling, involving informal waste collectors and recyclers. The MSW system uses mostly the sanitary landfill and incineration methods for waste treatment. Furthermore, the transition towards circularity of the value chain indicates that the country has introduced a number of general policies for waste management and a series of plastic policies to reduce, prohibit or ban some single-use plastics, address plastic pollution and improve the separation, collection, transportation, sorting and recycling of plastic waste. There are arguments, however, that a comprehensive regulatory framework for plastic packaging and recycling is lacking.

Germany is the leading producer of plastics in Europe, accounting for more than 20% of the region's plastics demand in 2023. The country is also a big consumer of the product. With the highest production in Europe, the country has a significant place in the region's plastics markets. The waste collection service in Germany is functional. A significant percentage of households in the country experience good waste collection service, as separate waste collection is practised extensively. The country generally has a culture of separating waste at source, resulting in increased recyclable recovery. The recycling sector is relatively established, forming an integral part of the sector. The transition towards circularity of the value chain reveals the various interventions introduced by the government to promote circularity of plastics, particularly to motivate consumers or households to separate their waste at source.

The United Kingdom is among the top plastic producers in the EU, with 3.3 million tonnes of plastic. It is highly dependent on polymer imports. About half of the polymer the country consumes annually is produced domestically, as a testament to the success of the sector. The waste management system is functional, with support provided by most local authorities for separate waste collection of recyclables. However, less than a third of the generated post-consumer plastic waste in the country is recycled. The transition towards circularity of the value chain indicates that the UK government continues to introduce general and specific plastic policies to address plastic waste, reduce single-use plastics consumption and advanc circularity of plastics.

References

africon, 2024. Distribution of Plastic Consumption in Nigeria in 2017, by Industry. Statista. Available at https://www.statista.com/statistics/1192675/plastic-consumption-in-nigeria-by-industry/.

Alves, B., 2025. Plastic waste in the UK – Statistics & facts. Statista. Available at https://www.statista.com/topics/4918/plastic-waste-in-the-united-kingdom-uk/.

Bardhan, A.R., Bhowmick, S., Mookherjee, P., 2024. Promoting circularity in the value chain of plastics in India. Observer Research Foundation. Available at https://www.orfonline.org/research/promoting-circularity-in-the-value-chain-of-plastics-in-india.

British Plastics Federation, 2025. About the British Plastics Industry. British Plastics Federation.

Chemonics Egypt, 2021. Study on Plastic Value Chain in Egypt. United Nations Industrial Development Organization. Available at https://www.unido.org/sites/default/files/unido-publications/2024-12/Plastic_value_chain_in_Egypt.pdf.

Confederation of Indian Industry, & World Wide Fund for Nature–India, 2022. India Plastics Pact: Roadmap to 2030. CII-ITC Centre of Excellence for Sustainable Development. Available at https://www.indiaplasticspact.org/website_assets/images/Roadmap-to-2030-1.pdf.

Department of Environment, Forestry and Fisheries, 2020. National waste management strategy 2020. Republic of South Africa. Available at https://www.dffe.gov.za/sites/default/files/docs/nationalwaste_management_strategy.pdf.

Drewniok, M.P., Gao, Y., Cullen, J.M., Cabrera Serrenho, A., 2023. What to do about plastics? Lessons from a Study of United Kingdom plastics flows. Environmental Science and Technology 57 (11), 4513–4521. Available at https://doi.org/10.1021/acs.est.3c00263.

Duru, R.U., Ikpeama, E.E., Ibekwe, J.A., 2019. Challenges and prospects of plastic waste management in Nigeria. Waste Disposal & Sustainable Energy 1 (2), 117–126. Available at 10.1007/s42768-019-00010-2.

EarthAction, 2024. Plastic Overshoot Day: 2024 report. Earth Action. 12.

Egypt Today Staff, 2025. 7 highlights of Egypt's 'Waste 50 by 2050' initiative in Africa. Egypt Today Staff. Available at https://www.egypttoday.com/Article/1/121577/7-highlights-of-Egypt-s-Waste-50-by-2050-initiative.

Elliot, T., Elliot, 2018. A plastic future plastics consumption and waste management in the UK. World Wide Fund for Nature. Eunomia Research and Consulting Ltd.

Emami, N., Baynes, T.M., Kaushik, T., Singh, M., Bhattacharjya, S., Locock, K., Schandl, H., 2024. Plastics in the Indian economy: a comprehensive material flow analysis. Journal of Material Cycles and Waste Management 26 (6), 3584–3595. Available at 10.1007/s10163-024-02060-z.

European Environmental Agency, 2022. Early warning assessment related to the 2025 targets for municipal waste and packaging waste: Germany. European Environmental Agency.

Ezeudu, O.B., Tenebe, I.T., Ujah, C.O., 2024. Status of Production, Consumption, and End-of-Life Waste Management of Plastic and Plastic Products in Nigeria: Prospects for Circular Plastics Economy. Sustainability 16 (18), 7900. Available at https://doi.org/10.3390/su16187900.

GPAP, 2024. Global Plastic Action Partnership. Nigerian National Plastic Action Partnership Towards a System change in plastic pollution: A roadmap for the sustainable use of plastics in Nigeria 12 https://weforum.ent.box.com/s/wcamitzoaom9vw4081t845z0myfp88dr

Greencape, 2019. Waste: 2019 market intelligence report.

Hahladakis, J.N., Velis, C.A., Weber, R., Iacovidou, E., Purnell, P., 2018. An overview of chemical additives present in plastics: Migration, release, fate and environmental impact during their use, disposal and recycling. Journal of Hazardous Materials 344, 179–199. Available at https://doi.org/10.1016/j.jhazmat.2017.10.014.

Hoffman, M., Schencks, C., 2020. The value chain and activities of polyethylene terephthalate plastics in the South African waste economy. Local Economy 35 (5), 523–535. Available at https://doi.org/10.1177/0269094220931697.

International Pollutants Elimination Network, & Shenzhen Zero Waste Environmental Public Welfare Undertakings Development Center, 2022. Plastic Waste Management and Burden in China. International Pollutants Elimination Network (IPEN). Available at https://ipen.org/documents/plastic-waste-management-and-burden-china.

Jaarsveldt, 2016. Waste management and recycling in South Africa Southern African – German Chamber of Commerce and Industry NPC.

Liang, H., Dong, H., Zhang, C., Geng, Y., Liu, X., Liu, G., Zhong, C., 2023. Combining LCA-MFA models to identify China's plastic value chain environmental impact mitigation pathways. iScience 26 (9), 107701. Available at https://doi.org/10.1016/j.isci.2023.107701.

Lindner, C., Schmitt, J., Hein, J., 2019. Conversio Market & Strategy GmbH Material flow analysis plastics in Germany https://www.conversio-gmbh.com/res/News_Media/2020/Summary_Material_Flow_Analysis_Plastics_Germany_2019_EN.pdf

Liu, C., Liu, C., 2023. Exploring plastic-management policy in China: status, challenges and policy insights. Sustainability 15 (11), 9087. Available at https://doi.org/10.3390/su15119087.

Mofo, L., 2020. Future-proofing the plastics value chain in Southern Africa WIDER Working Paper 2020/148. Available at https://doi.org/10.35188/UNU-WIDER/2020/905-1.

Nassour, A., Hemidat, S., Lemke, A., Elnaas, A., Nelles, M., 2018. Separation by manual sorting at home: State of the art in Germany. Handbook of Environmental Chemistry 63, 67–87. Available at https://doi.org/10.1007/698_2017_26.

Nikam, J., Ddiba, D., Njoroge, G., 2022. Analysis of the plastic waste value chain in India: A scoping study. Stockholm Environment Institute. Available at 10.51414/sei2022.037.

Nzeadibe, T.C., Ajaero, C.K., 2011. Development Impact of Advocacy Initiatives in Solid Waste Management in Nigeria. Environment, Development and Sustainability 13 (1), 163–177. Available at https://doi.org/10.1007/s10668-010-9254-8.

Obrecht, M., Haddad, R.E., Elbary, R.A., Lukman, R.K., Rosi, M., 2019. Sciendo Slovenia PROMOTING SUSTAINABLE AND CIRCULAR PLASTICS USE IN EGIPT WITH IMPLEMENTATION OF ECODESIGN PRINCIPLES System Safety: Human - Technical Facility - Environment 1 1 441-448 https://sciendo.com/journal/CZOTO 10.2478/czoto-2019-0057

Olatayo, K.I., Mativenga, P.T., Marnewick, A.L., 2024b. Pathways to zero plastic waste landfill: Progress toward plastic circularity in South Africa. Journal of Industrial Ecology 28 (5), 1270–1288. Available at https://doi.org/10.1111/jiec.13533.

Olatayo, K.I., Mativenga, P.T., Marnewick, A.L., 2024a. Measuring the performance and maturity of the plastic recycling value chain system: implications and prospects. Environmental Sciences Europe 36 (112). Available at https://doi.org/10.1186/s12302-024-00937-2.

Oluwadipe, S., Garelick, H., McCarthy, S., Purchase, D., 2022. A critical review of household recycling barriers in the United Kingdom. Waste Management and Research 40 (7), 905–918. Available at https://doi.org/10.1177/0734242X211060619.

Picuno, C., Alassali, A., Chong, Z.K., Kuchta, K., 2021. Flows of post-consumer plastic packaging in Germany: An MFA-aided case study. Resources, Conservation and Recycling 169, 105515. Available at https://doi.org/10.1016/j.resconrec.2021.105515.

Plastics SA, 2017. Plastics SA annual report 2016/2017. Plastics SA. Available at https://www.plasticsinfo.co.za/wp-content/uploads/2019/09/PlasticsSA-annual-report-20167-A5-email.pdf.

Plastics SA, 2022. Plastics 2022: An analysis of the South African plastics industry data. Plastics SA. Available at https://www.plasticsinfo.co.za/wp-content/uploads/2023/12/PlasticsSA-Annual-Review-20222023_interactive.pdf.

PwC, 2019. Market assessment for plastics circularity in Nigeria [Draft report]. West Africa Coastal Areas Management Program. Available at https://www.wacaprogram.org/knowledge/market-assessment-plastics-circularity-nigeria.

Shanker, R., Khan, D., Hossain, R., Islam, M.T., Locock, K., Ghose, A., Sahajwalla, V., Schandl, H., Dhodapkar, R., 2023. Plastic waste recycling: existing Indian scenario and future opportunities. International Journal of Environmental Science and Technology 20 (5), 5895–5912. Available at https://doi.org/10.1007/s13762-022-04079-x.

Singh, S., 2020. Decentralized solid waste management in India: A perspective on technological options. National Institute of Urban Affairs.

South African Plastics Pact, 2021. South African Plastics Pact: 2025 Targets. South African Plastics Pact. Available at https://www.saplasticspact.org.za/2025-targets/.

Statista Research Department, 2025. Plastic consumption in Germany in 2021 and 2023. Statista. Available at https://www.statista.com/statistics/863436/plastic-consumption-in-germany/.

United Nations Environment Programme, 2025. UNEP Plastics Initiative. United Nations Environment Programme. Available at https://www.unep.org/topics/plastic-pollution/one-plastics-initiative/about-unep-plastics-initiative.

Viljoen, J., Blaauw, D., Schenck, C., 2019. The opportunities and value-adding activities of buy-back centres in South Africa's recycling industry: A value chain analysis. Local Economy 34 (3), 294–315. Available at https://doi.org/10.1177/0269094219851491.

Waste and Resources Action Programme, 2020. European Plastics Pact Roadmap. WRAP. Available at https://www.wrap.ngo/taking-action/plastic-packaging/initiatives/the-uk-plastics-pact.

Weiner, K., Butler, J.K., Neumann, A., 2025. Global Compliance News Germany: Circularity made in Germany – The national circular economy strategy has been adopted 12 https://www.globalcompliancenews.com/2025/01/24/https-insightplus-bakermckenzie-com-bm-environment-climate-change_1-germany-circularity-made-in-germany-the-national-circular-economy-strategy-has-been-adopted_01172025/

World Bank, 2022. Nigeria Plastic Country Brief. The West and Central Africa Programme.

Wrap, 2024. The UK plastics pact. Waste and Resources Action Programme 12.

Xiao, S., Dong, H., Geng, Y., Brander, M., 2018. An overview of China's recyclable waste recycling and recommendations for integrated solutions. Resources, Conservation and Recycling 134, 112–120. Available at https://doi.org/10.1016/j.resconrec.2018.02.032.

Yee, A.S., 2023. Managing Plastic Waste in the People's Republic of China. ADB Briefs 249.

Zhang, D.Q., Tan, S.K., Gersberg, R.M., 2010. Municipal solid waste management in China: Status, problems and challenges. Journal of Environmental Management 91 (8), 1623–1633. Available at https://doi.org/10.1016/j.jenvman.2010.03.012.

Chapter 11

Conclusion

11.1 Book conclusion

This book has shown both the indispensable role of plastic materials and products in modern economies and the mounting global crisis created by plastic waste. The circular economy as a viable path forward was underscored; it is one that demands transformative action at every level and across every stage of the plastics value chain. An understanding of the plastics value chain and the critical role each value chain stage plays is important for informed decisions on implementing circular economy practices, driving forward circular economy businesses and addressing or avoiding environmental impacts from plastic waste.

In advancing a circular economy for plastics, options higher up the waste management hierarchy are advocated as they help retain higher value and minimise the environmental impact of circular economy processing. The evolution of the waste management hierarchy, from the Waste Framework Directive 2008/98/EC in 2008, to the amendments and then on to the directive in 2018, has promoted a priority order from the best desirable option of prevention at the top of the hierarchy to the least desirable option of disposal at the bottom. Further analysis of the EU waste hierarchy resulted in the development of the R framework systems. The more comprehensive R framework covers refuse, rethink, reduce, reuse, repair, refurbish, re-manufacture, repurpose, recycle and recover. Therefore countries, stakeholders and value chain actors design and implement plastics management options, promoting and enabling a shift towards options higher up the waste management hierarchy.

Life cycle assessment is essential in quantifying plastic's impact on the environment. It should support sustainability assessment and consider the economic, social and environmental impacts and benefits. Economic sustainability contributes to long-term economic growth without impacting the environment negatively. Social sustainability promotes social benefits, for example equity, public participation, education, inclusivity and social cohesion, in accessing plastic resources. Environmental sustainability ensures resource and energy efficiency and reduced environmental impacts.

Plastic Circularity. DOI: https://doi.org/10.1016/B978-0-443-40667-6.00014-7

The transition to circularity needs to be supported by modelling material flows to enable visibility in flows and to inform circularity metrics. Data are pivotal in modelling material flow and circularity. There are several challenges in collecting plastic data across the value chain. These challenges limit the quality of the data in terms of accuracy, certainty, credibility and reliability. In addressing these limitations, a framework for building plastic data capacity in the value chain is important. Four data build-up processes, stages or datasets are required at the production, consumption, waste generation and management, recovery and recycling stages. In addition, national, regional and municipal registries for plastics help to provide unified governance, and a cohesive system and framework for the collection of plastic data to support a comprehensive dataset that can be used for resource flow and circularity modelling.

Material flow analysis is a tool for modelling, quantification and visualisation of mass flow and circularity performance across the value chain. A number of policy interventions are being introduced by national governments and international organisations to promote circularity at different phases of the plastics life cycle. These include deposit refund systems, tax refunds, mandatory utilisation, extended producer responsibility, plastic bag regulation, user pay schemes, landfill taxes/bans and kerbside collection systems for waste separation and management.

The effective participation of stakeholders in the plastics value chain is another critical aspect of transitioning the plastic system from linearity to circularity. The plastic system is dynamic and multi-staged, with multiple stakeholders in every stage having individual objectives that need to be coordinated. Therefore, value chain coordination becomes imperative, and this requires the design and development of a participatory framework of all the stakeholders for functional collaboration, insightful research, data sharing, information dissemination, efficient resource utilisation and waste resource accountability and traceability. Benefits associated with plastic systems with well-coordinated value chains include cost-efficiency, increased household recycling awareness and motivation, better waste handling and sorting, improved regulation enforcement and reduced expenses from cost sharing of facilities and infrastructure.

The pathway to zero landfill for plastics integrates relevant circularity interventions and practices that centre on reducing or eliminating plastic waste through sustainable plastic product design, substituting plastics with bio-based plastic or alternative material, applying the reuse model (reuse, repair and repurpose) and recycling and recovery. Similarly, the pathway to recycling plastic waste requires the recyclability of plastics to be optimised from product design to reprocessing, comprising strategies such as innovatively and sustainably designing products for recycling purposes, improving the collection and sorting of plastic waste for recycling and increasing recycling capacity.

The maturity model was presented to evaluate the level of development or maturity of any plastic recycling system, along five levels of maturity in increasing order of improvement. They are level 1 (ad hoc and unstructured), level 2 (structured), level 3 (visionary), level 4 (integrated) and level 5 (connected and dynamic). The maturity levels are guided by seven maturity elements for the assessment of the system: chain actors' integration, customer satisfaction, performance targets, process costs, quality data and database, quality recyclables and products, structured processes and system performance review. The assessment of the performance and development of the plastic system helps in the design of action plans to bridge the gaps between the current level of performance and maturity and the desired optimal system performance and full maturity.

The book ends with a review of how seven countries, namely South Africa, Nigeria, Egypt, India, China, Germany and the United Kingdom, are navigating the shift towards a circular plastics economy. Their value chain systems, policy frameworks and circularity goals were compared.

South Africa has introduced several circularity policies, yet enforcement remains a major barrier. Nigeria has no dedicated plastics circularity legislation, leaving the transition largely unattended. Egypt has begun reforming its waste system, but still needs stronger, system-wide interventions. India has launched national and state-level policies, though regulation is young and enforcement uneven. China has issued multiple plastic policies, but there is room for a fully coherent framework for packaging and recycling. Germany shows the most advanced transition, driven by strong incentives for household waste separation. The United Kingdom continues to expand policies targeting plastic waste, single-use reduction and circular design.

Together, these systems reveal a world in transition towards circularity at different paces. The journey towards circular plastics is far from complete, yet the direction is clear. The next chapter is shaped by the choices we make now.

Index

Note: Page numbers followed by "*f*" and "*t*" refer to figures and tables, respectively.

D

E

F

G

H

I

K

L

M

N

P

W

Z

Printed and bound by CPI Group (UK) Ltd, Croydon, CR0 4YY
24/09/2026
02222289-0003